Revolutionary Science

The Struggle for Agroecology in the Americas

BRUCE H. JENNINGS

critical development studies

Copyright © 2026 Bruce H. Jennings

All rights reserved. No part of this book may be reproduced or transmitted in any form by any means without permission in writing from the publisher, except by a reviewer, who may quote brief passages in a review. The publisher expressly prohibits the use of this work in connection with the development of any software program, including, without limitation, training a machine learning or generative artificial intelligence (AI) system

Copyediting: Jennifer Harris
Cover Design: John van der Woude
Text Design: Lauren Jeanneau
Printed and bound in Canada

Published in North America by Fernwood Publishing
2970 Oxford Street, Halifax, Nova Scotia, B3L 2W4
Halifax and Winnipeg
www.fernwoodpublishing.ca

Published in the rest of the world by Practical Action Publishing
27a Albert Street, Rugby, Warwickshire CV21 2SG, UK

Fernwood Publishing Company Limited gratefully acknowledges the financial support of the Government of Canada through the Canada Book Fund and the Canada Council for the Arts. We acknowledge the Province of Manitoba for support through the Manitoba Publishers Marketing Assistance Program and the Book Publishing Tax Credit. We acknowledge the Nova Scotia Department of Communities, Culture and Heritage for support through the Publishers Assistance Fund.

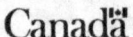 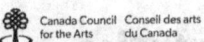 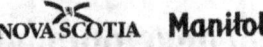

Library and Archives Canada Cataloguing in Publication
Title: Revolutionary science : the struggle for agroecology in the Americas /
by Bruce H. Jennings.
Names: Jennings, Bruce H., 1951- author
Description: Includes bibliographical references and index.
Identifiers: Canadiana 20250311607 | ISBN 9781773638065 (softcover)
Subjects: LCSH: Agricultural ecology—Latin America. | LCSH: Agricultural innovations—Latin America. | LCSH: Agriculture—Social aspects—Latin America.
Classification: LCC S473.9 .J46 2026 | DDC 630.98—dc23

ALSO IN THE CRITICAL DEVELOPMENT STUDIES SERIES

Lessons from the Zapitistas: From Armed Insurgency to People's Autonomy (2025)
by Lia Pinheiro Barbosa and Peter M. Rosset

Hidden Politics in the UN Sustainable Development Goals (2024)
by Adam Sneyd

Contested Global Governance Space and Transnational Agrarian Movements (2023)
by Mauro Conti

The Political Economy of Agribusiness (2023)
by Maria Luisa Mendonça

Global Fishers: The Politics of Transnational Movements (2023)
by Elyse Noble Mills

Tiny Engines of Abundance (2022)
by Jim Handy

COVID-19 and the Future of Capitalism: Postcapitalist Horizons Beyond Neo-Liberalism (2021)
by Efe Can Gürcan, Ömer Ersin Kahraman & Selen Yanmaz

Extractivism: Politics, Economy and Ecology (2021)
by Eduardo Gudynas

The Political Economy of Agrarian Extractivism: Lessons from Bolivia (2020)
by Ben M. McKay

Development in Latin America: Toward a New Future (2019)
by Maristella Svampa, translation by Mark Rushton

Politics Rules: Power, Globalization and Development (2019)
by Adam Sneyd

Critical Development Studies: An Introduction (2018)
by Henry Veltmeyer & Raúl Delgado Wise

Contents

Critical Development Studies Series vi
Series Editors vii
Glossary viii
Acronyms and Abbreviations x
Acknowledgements xii
Introduction 1

Part I — Conflicts in Science: Revolution and Counterrevolution in the Americas / 7

1 Scientific and Social Revolutions 8
   Industrial Agriculture versus Agroecology 9
   The Rockefeller Mission in Mexico 13
   Notes 14

2 Transforming Mexican Agriculture 16
   A Science for Whom? 21

3 The Counterrevolution in Mexican Agriculture 27

4 The Production of Knowledge 34
   The Issue of Alternatives in Science 40
   Science and Agency 41
   Notes 44

5 Expanding a Green Revolution across the Americas 45
   Exporting the Model: From Mexico to the Americas 45
   Training a Generation of Técnicos 48
   Notes 52

6 The Violence of the Green Revolution 53
   Beyond Mexico 58
   Creative Destruction 60
   Notes 63

Part II — Critiquing a Dominant Science and Its Larger Consequences / 65

7 Ecology, Chemistry, and Conflicts 66
   Industrial Agriculture: Conflicts over Chemical Consequences 68
   Note 72

8 Fights in the Fields 73
   Industrial Agriculture and California's Poisoned Workers 74
   Fields of Knowledge and Conflicting Claims 76
   Scientific Reductionism and Its Political Consequences 77
   Notes 79

9	Agrichemicals and the Law	80
	Big Green	82
	Hiding the Evidence of Harm	83

Part III — The Struggles of a New Science / 87

10	An Alternative in Production	88
	Producing an Alternative	92
	Notes	94
11	Demonstrating Another Knowledge and Practice	95
	Fixing Industrial Agriculture	97
	A Gathering of Other Scientists	98
	Note	100
12	New Markets / New Conflicts	101
	Agricultural Technologies, Free Trade, and Structural Violence	105
	International Agricultural Research Centres and Agroecology	107
	Note	109
13	Agroecology at Berkeley: A Path Not Taken	110
	Challenging New Projects in Science and Technology	111
	Notes	117

Part IV — Conclusions:
Linking a Science and a Movement / 119

14	Agroecology and Infrastructures of Support	120
	Agroecological Lighthouses	122
	Agroecology and Community Restoration of Watersheds	123
	Agroecology and Pedagogical Infrastructure	125
	Agroecological Infrastructures of Food Sovereignty	127
	Agroecology and Alternative Market Infrastructures	128
	The Limits of Supportive Infrastructures	131
	Note	132
15	Agroecology and Infrastructures of Resistance	133
	The Political Economy of Ignorance: Industrial Agriculture and Its Consequences	135
	Joining and Creating Movements of Political Resistance	136
	Expanding Spaces for the Exercise of Democratic Rights	138
	Notes	140
Epilogue		141
	Notes	145
References		146
The Struggle Continues		153
Index		154

Critical Development Studies Series

Three decades of uneven capitalist development and neoliberal globalization have devastated the economies, societies, livelihoods and lives of people around the world, especially those in societies of the Global South. Now more than ever, there is a need for a more critical, proactive approach to the study of global and development studies. The challenge of advancing and disseminating such an approach — to provide global and development studies with a critical edge — is on the agenda of scholars and activists from across Canada and the world and those who share the concern and interest in effecting progressive change for a better world.

This series provides a forum for the publication of small books in the interdisciplinary field of critical development studies — to generate knowledge and ideas about transformative change and alternative development. The editors of the series welcome the submission of original manuscripts that focus on issues of concern to the growing worldwide community of activist scholars in this field. Critical development studies (CDS) encompasses a broad array of issues ranging from the sustainability of the environment and livelihoods, the political economy and sociology of social inequality, alternative models of local and community-based development, the land and resource-grabbing dynamics of extractive capital, the subnational and global dynamics of political and economic power, and the forces of social change and resistance, as well as the contours of contemporary struggles against the destructive operations and ravages of capitalism and imperialism in the twenty-first century.

The books in the series are designed to be accessible to an activist readership as well as the academic community. The intent is to publish a series of small books (54,000 words, including bibliography, endnotes, index and front matter) on some of the biggest issues in the interdisciplinary field of critical development studies. To this end, activist scholars from across the world in the field of development studies and related academic disciplines are invited to submit a proposal or the draft of a book that conforms to the stated aim of the series. The editors will consider the submission of complete manuscripts within the 54,000-word limit. Potential authors are encouraged to submit a proposal that includes a rationale and short synopsis of the book, an outline of proposed chapters, one or two sample chapters, and a brief biography of the author(s).

Series Editors

HENRY VELTMEYER is a research professor at Universidad Autónoma de Zacatecas (Mexico) and professor emeritus of International Development Studies at Saint Mary's University (Canada), with a specialized interest in Latin American development. He is also co-chair of the Critical Development Studies Network and a co-editor of Fernwood's Agrarian Change and Peasant Studies series. The CDS *Handbook: Tools for Change* (Fernwood, 2011) was published in French by University of Ottawa Press as *Des outils pour le changement : Une approche critique en études du développement* and in Spanish as *Herramientas para el Cambio*, with funding from Oxfam UK by CIDES, Universidad Mayor de San Andrés, La Paz, Bolivia.

ANNETTE AURÉLIE DESMARAIS is the Canada Research Chair in Human Rights, Social Justice and Food Sovereignty at the University of Manitoba (Canada). She is the author of *La Vía Campesina: Globalization and the Power of Peasants* (Fernwood, 2007), which has been republished in French, Spanish, Korean, Italian and Portuguese, and *Frontline Farmers: How the National Farmers Union Resists Agribusiness and Creates our New Food Future* (Fernwood, 2019). She is co-editor of *Food Sovereignty: Reconnecting Food, Nature and Community* (Fernwood, 2010); *Food Sovereignty in Canada: Creating Just and Sustainable Food Systems* (Fernwood, 2011); and *Public Policies for Food Sovereignty: Social Movements and the State* (Routledge, 2017).

RAÚL DELGADO WISE is a research professor and director of the PhD program in Development Studies at the Universidad Autónoma de Zacatecas (Mexico). He holds the prestigious UNESCO Chair on Migration and Development and is executive director of the International Migration and Development Network, as well as author and editor of some twenty books and more than a hundred essays. He is a member of the Mexican Academy of Sciences and editor of the book series, Latin America and the New World Order, for Miguel Angel Porrúa publishers and chief editor of the journal *Migración y Desarrollo*. He is also a member of the international working group, People's Global Action on Migration Development and Human Rights.

Glossary

ahupuaʻa: Hawaiian term for a portion of land, often shaped like a pie slice, extending from the top of a valley or mountain to the ocean

ʻāina: Hawaiian term referring to land that reflects its historical, ecological, social and spiritual values

Article 27: a part of the Mexican Constitution relating to property rights, including provisions establishing communal land tenure to groups of campesinos. Until 1992, it prohibited the sale of ejidos and communal land

campesino/a: Spanish term meaning peasant; as used in this book, generally refers to people engaged in a range of agricultural pursuits — partially employed workers on small plots and farms, fisherfolk, as well as landless workers in peri-urban agricultural industries and regional markets

campesino a campesino: social methodology based on the collective participation and central role of campesinos guiding decisions about agricultural production, including agroecological transformations of their territories

chinampas: raised beds surrounded by canals that deliver water to the root zone of plants; a traditional agricultural practice extending over thousands of acres that provided sustenance to the people of Tenochtitlán (present-day Mexico City)

diálogo de saberes: dialogue bridging distinct kinds of knowledge with different origins, containing no preconceived hierarchy or superiority

ejidatarios: users of ejido land

ejido: A land grant, typically established after the Mexican Revolution, operated collectively by a group of members

extensionismo: as used here, a semi-pejorative term referring to extension workers having expertise in industrial agriculture

hacendados: owners of haciendas

haciendas: colonial estates, many of which were formed prior to the Mexican Revolution, frequently representing very large landholdings

indígena: Indigenous person, representing one of the many hundreds of Indigenous communities across the Americas

maíz: maize or corn

patronato: literally "patron"; here refers to a wealthy landholder or association of employers; a group increasingly linked to larger regional, national, or even international financial interests

técnicos: agricultural extension workers

tianguis (from Nahuatl): a local or regional market

transgénico: transgenic; generally a shortened reference to a genetically engineered or modified plant

waru-waru: traditional agricultural practice based on raised planting beds surrounded by canals for flood and frost protection, improved soil fertility, and enhanced crop yields.

Zapatistas: a group of campesinos and others who led a national insurrection that began in Chiapas to protest neoliberal policies advancing free trade as well as the destructive character of capitalism across Mexico

Acronyms and Abbreviations

CGIAR	Consultative Group on International Agricultural Research
CIAT	Centro Internacional de Agricultura Tropical, International Centre for Tropical Agriculture (Colombia)
CIMMYT	Centro Internacional de Mejoramiento de Maíz y Trigo, International Maize and Wheat Improvement Center (Mexico)
CIOAC	La Central Independiente de Obreros Agrícolas y Campesinos, Independent Centre of Agricultural Workers and Campesinos
CIP	El Centro Internacional de la Papa, International Potato Centre (Peru)
CLADES	Consorcio Latinoamericano sobre Agroecología y Desarrollo Latin American Consortium for Agroecology and Development
ECOSUR	El Colegio de la Frontera Sur, College of the Southern Frontier
FIOAC	Federación Independiente de Obreros Agricolos y Campesinos, Independent Federation of Agricultural Workers and Campesinos
GEB	General Education Board, Rockefeller Foundation
IALA	Instituto Agroecológico Latinoamericano, Latin American Institute of Agroecology
IIA	Instituto de Investigaciones Agrícolas, Institute for Agricultural Investigations
INIA	National Institute for Agricultural Investigations
IRRI	International Rice Research Institute

LVC	La Vía Campesina
MAP	Mexican Agricultural Program, Also referred to as OEE/OSS
NAFTA/TLC	North American Free Trade Agreement, Tratado de Libre Comercio (Tratado)
NFU	National Farmers Union (Canada)
OEE/OSS	Oficina de Estudios Especiales, Office of Special Studies (MAP)
OTA	Office of Technology Assessment
SOCLA	Sociedad Científica Latinoamericana de Agroecología, Latin American Scientific Society of Agroecology
UACH	Universidad Autónoma de Chapingo, Autonomous University of Chapingo
UC	University of California
UGOCM	Unión General de Obreros y Campesinos de México, General Union of Mexican Workers and Campesinos
USDA	United States Department of Agriculture

Acknowledgements

Among the many persons who share responsibility for the production of this book, Cheri Lucas Jennings has provided multiple roles: a decades-long companion on this journey, an influence on the structure of this book, and as an intellectual guide in constructing this history. It is particularly this last role in which Cheri has provided an empathic yet critical perspective on science and scientists. In addition to drawing on Cheri's voice throughout this book, her presence and perspectives are celebrated in the Epilogue.

I am especially grateful for a group of colleagues and others who, with little advance notice, graciously provided invaluable commentary and guidance on drafts of this book. These readers included Jill Belsky, Steve Siebert, Joel Tickner, and Ben Kerkvliet — a group of individuals who have assisted me in innumerable ways over many years with a wealth of challenging and insightful conversations. My gratitude as well to members of my Mānoa 'Ohana, including Michael Shapiro, Kathie Kane, and Noel Kent.

There are also many of my colleagues working in or around the California Legislature who have inspired facets of my writing including Ralph Lightstone and his colleagues with the California Rural Legal Assistance Foundation, Tom Rankin, along with many others of my brothers and sisters in the labour movement, colleagues with the Senate Office of Research, and a very large contingent of advocates representing farmworkers, communities facing toxic hazards, industrial and chemical workers, and many others across the state. I would be remiss without mentioning Martha Guzman Aceves and her association with the United Farm Workers, including Dolores Huerta, as providing me with a continuing inspiration to expand public conversations regarding the struggles of farmworkers in California and beyond.

Among the most important contributors to this work are the many Latin American agroecologists who have given much of their time to educate me about the many facets of agroecology. This group includes Clara Nicholls, Juan Sanchez Barba, Walter Pengue, Agustin Infante, Rolando Rojas, Santiago Peredo Parada, Claudia Barrera-Salas, Mauricio Chang, and Rene Montalba. In addition to these people are many others associated with agroecological projects who have assisted me during my research.

Miguel Altieri and Peter Rosset are two of the leading figures in this book. Miguel's generosity has extended to inviting me to accompany him on travels, to join him at conferences, to provide introductions to many of his colleagues across the Americas, and to engage with a series of tutorials in recent years. I am similarly indebted to Peter Rosset, who has also played a major role in shaping my understanding of the political dimensions of agroecology. I would encourage any readers wishing to delve more deeply into the science and politics of agroecology to watch any of the many videos that Miguel and Peter have made available on various platforms.

I also want to direct readers to a vitally important group of agroecologistas whose written, video, and other presentations have informed my writing, including Clara Nicholls, Helda Morales, Juliana Sabugal, Marina Sanchez de Praeger, Laura M. Fuertes Sanchez, Paula Andrea Rugeles, Angela Maria Londono Matta, Gloria Guzman Casado, Georgina Catacora-Vargas, Irene Cardoso, and Marta Soler-Montiel. These individuals constitute only a few of the many Latin American women whose contributions underscore the pivotal importance of agroecology as an expansive and inclusive social movement and scientific undertaking.

While only recently acquainted with my work, Colleen Ford Meseroll has played a vital role in encouraging me to complete this writing project in a timely fashion (i.e., before it becomes memorialized as a posthumous publication). In addition to indicating how various passages might be improved with an occasional verb or noun, she has provided refuge with her laughter, culinary artistry, and even an occasional Manhattan.

There are also several individuals who, although no longer alive, continue to have a very large presence in my life, my writings, and my conduct. These individuals begin with Donald Dahlsten and Edmund K. Oasa — and Ernest Feder, who arranged for our earliest collaborative work in Mexico's El Trimestre Economico. Bill Friedland, a professor at UC Santa Cruz was also among the earliest and most encouraging voices on the investigations Ed Oasa and I conducted independently of but collaboratively with IRRI in the Philippines and CIMMYT in Mexico. I also wish to acknowledge the productive arguments Cheri and I frequently had with Wes Jackson, Donald Worster, Angus Wright, and John Perkins.

My special thanks goes to the multitalented editor and collaborator, Iris J. Thuesen. Iris has provided constant and superlative assistance as one of the primary sources of support for the writing, editing, and production of this work. Working from Buenos Aires, Iris has been instrumental in making this work more accessible to a broader audience while enriching the essential themes. The fact that she has performed all manner of support for this writing project while creating a new household for Franco as well as newly arrived Kai and Apollonia Cheri is truly impressive.

A final word of gratitude goes to Fernwood Publishing's founder, Errol Sharpe, and Fernwood's talented team, including Lauren Jeanneau and Anumeha Gokhale. Mr. Sharpe provided insightful and patient guidance coupled with good humour, for which I am extremely grateful. I have an even deeper appreciation for what Fernwood Publishing provides as a vanishing commitment in the world: publishing a collection of works dedicated to providing critical writing for critical minds.

Introduction

It was a place unlike any other I had experienced in my young life. I awoke to the pounding of the ocean surf. The campsite was at the mouth of a valley where a waterfall delivered a seemingly endless supply of crystal clear water. It was a lush valley with a diversity of plants and life. And all of this appearing on an island in the middle of an ocean. It was obvious why so many people referred to this place as paradise.

In that moment, however, my attention was focused on the quite lovely person next me; a young woman who would soon become my partner for decades to come. Cheri Lucas Jennings then distracted me into thinking about breakfast and what to add to the meagre provisions we had brought from Honolulu to Hanakāpīʻai, a now uninhabited valley along the Nāpali Coast on the island of Kauaʻi in the state of Hawaiʻi.

After spending the night in this remote valley, Cheri searched for and found herbs to add to our breakfast. Her discovery revealed plants emerging from a terraced stonework reflecting a system of irrigation that had endured for many decades. All of this suggested a system of food production that meshed perfectly with the surrounding ecology. The enduring architecture of this system would, in years to come, remind us of the genius of the early Hawaiians who once populated these islands in great numbers.

Within a couple of years, Cheri and I would relocate our doctoral research to Kauaʻi, a place where we would live at the back of another valley and not far from where Cheri taught at the local community college. It was here where Cheri and I witnessed firsthand more traces of the early Hawaiians' astonishing construction of a system where intensive agricultural production was integrated into divisions of land beginning high in the mountains, continuing through forests and meadows, and terraces typically ending with streams feeding nutrients into fish ponds. These extraordinarily complex systems — an ecologically based architecture of living systems that the Hawaiians referred to as ahupuaʻa[1] — were at that time only beginning to be more fully recognized and understood by non-natives in this place.

In striking contrast to the ingenuity of ancient Hawaiian systems of blending agriculture and the surrounding ecology stood the sugar cane and pineapple plantations that still dominated much of Hawai'i's landscapes in the 1970s. The sheer brutality of plantation agriculture was evident in the treatment of the land and waters, disregarding any sense of the surrounding ecology. The design of plantations cut across and through any existing ahupua'a with systems of production focused solely on yields of a single commodity for export to distant lands. The overriding goal of production was antithetical to the deeply embedded concept of 'āina held by Hawaiians. From the perspective of Hawaiians, 'āina conveyed a sense of land as a sacred entity, with landscapes deserving care and respect. It was not simply that plantations resulted in damaging environmental impacts; they fostered a broader array of negative consequences, including the community's access to a diversity of plants for medicines and food, jobs linked to maintaining ahupua'a, education centred on understanding the entirety of island ecosystems, and a system of governance ensuring that the uses of the land were integrally linked to providing food for the island's community. In a place with such tremendous potential for both feeding a substantial population and preserving its natural ecology, the legacy of plantations meant that access to land and food would become increasingly precarious for the residents of these islands.

Our experience at Hanakāpī'ai, introductions to the world of a'ina and the multiple dimensions of ancient landscapes, along with reflections on plantations and their larger consequences would slowly develop into deeper investigations into the science of industrial agriculture. The seed was planted in our minds that something was profoundly flawed with the kind of agriculture taking hold in many parts of the world. Our research evolved into much longer discussions about the rationale explaining how the ancient agricultural systems that had flourished throughout these islands could be so thoroughly and rapidly dismantled and replaced by industrial agriculture; how paradise could be so thoroughly displaced.

The islands' history of colonization by foreign powers was accompanied by a thorough replacement of its abundance of a diverse source of local foods into what would become a largely industrial model of agriculture for the export of sugar and pineapples. Long before the close of the twentieth century, the islands were importing upwards of 90 percent of its foods from the United States mainland. With the vanishing of local food production and the ever-increasing prices for imported foods, residents of the islands wondered where all of this would end. What, if anything, could they do to achieve some kind of alternative? Was it possible to recapture those aspects of life that made this place a paradise on Earth?

Coinciding with our lives on Kaua'i, our research increasingly led us back to a less recognized set of institutions that had been so important to transforming agriculture on a global scale. These institutions — from agricultural colleges to international agricultural research centres — frequently figured prominently in defining the design of agricultural production throughout the world. More than anything else, this may explain why our community of mentors and colleagues encouraged us to examine science itself, a process many of us refer to as the "politics of science." In the years that followed, we gained a greater appreciation that dominant narratives of selected scientific projects have severely misrepresented these works in the world. More specifically, what is generally received as agricultural science is influenced by significant political considerations.

Coinciding with the time we were initiating our studies, a young Latin American was engaging in a separate investigation of ancient agricultural systems and the scientific complexity of the multiple layers of benefits purposely designed into these systems. This young scientist, Miguel Altieri, would bring attention to what would become a recognized alternative to industrial agriculture: agroecology. The work of Miguel, his colleagues, and innumerable agricultural workers across the Americas would eventually become recognized as a competing science, but also as a social movement dedicated to advancing agroecology.

One of the most concise definitions of agroecology, particularly for readers in Canada and the United States, can be found in the statements written approximately a decade ago by the National Farmers Union (NFU), an organization representing many thousands of Canadian farmers and farmworkers. To be clear, the NFU statement largely paraphrases earlier definitions authored by a group of Latin American agroecologists as a guide regarding its own struggles in Canada.

The NFU acknowledges that many of the scientific dimensions of agroecology are anchored in traditional agricultural practices devised by Indigenous people and communities across the Americas. Many of these practices, often honed over centuries, have evolved into agricultural systems that maintain, or even enhance, the ecology of surrounding landscapes. For scientists who have studied these traditions for decades, especially Latin America's agroecologists, the practices still used by many Indigenous and small farmers throughout the America's represent a wealth of knowledge about how to devise an agriculture yielding a much broader array of benefits than the industrial agricultural systems that dominate landscapes in the United States and Canada. The statement issued by the Canadian farmers and farmworkers is especially striking in how it situates agroecology and agroecologists in the context of an epic struggle between the forces supporting industrial agriculture versus the hundreds of millions of campesinos and agroecologists who are advancing a very different vision for the future.

Miguel Altieri figures prominently in this book, first as a notable agroecologist whose work has included numerous professional activities across North and South America. Indeed, Miguel Altieri is generally regarded as one of the most recognized authorities on agroecology. Readers looking for a more comprehensive and thorough explanation about agroecology's principles and practices are strongly encouraged to consult the seminal work authored by Miguel Altieri and Peter Rosset (Rosset and Altieri 2017).

Dr. Altieri's position as a professor at the University of California–Berkeley has intersected at various points with my own career as a researcher and policy advisor for the state's premier lawmaking body, the California Legislature. My responsibilities to protect farmworkers and their communities from agrichemicals and other toxic substances often placed me in an adversarial position with lobbyists representing petroleum and chemical companies, agribusinesses, and major financial interests; whereas I represented the people of California, these corporate lobbyists were typically accountable to a very small group of wealthy and powerful individuals.

It is important for readers to recognize that my accomplishments, like Miguel's, have involved a much larger community of scholars, activists, and ordinary workers. This book seeks to highlight the vital and pivotal role of that larger collective, ensuring that any accomplishments discussed are not attributed to individuals alone but understood as the countless efforts by many others. Whereas my previous published work revealed the political dimensions of the international agricultural research centres, this work expands on that history with the study of a newly emergent science. The new science of agroecology is noteworthy as it represents an alternative to the agricultural sciences of recent decades and their emphasis on industrial agriculture. One of agroecology's most distinctive characteristics is its explicit link to the traditional knowledge of those who work the land — campesinos — a group referring to farmworkers but also many others. Beyond this trait, leading voices of agroecology across the Americas insist that their work is not just about studying agricultural systems, but about advancing the role of campesinos in shaping the field itself. Integrating this social purpose is what reveals the revolutionary potential of this new science.

Casual observers may argue that the social purpose of agroecology means it has moved beyond the realm of science. A narrative that demarcates a divide between science and politics, however, is deeply flawed. Our work documents how the importing of the agricultural sciences from the United States to the rest of the Americas aligned with a very specific set of social and political interests. In that same vein, we focus on what the current struggle for agroecology means for people across the Americas and particularly for those living in the United States.

The loss of Hawai'i's paradise is not unique. As I earlier described our initial witnessing of the losses accompanying the rise of plantations in Hawai'i, we would grow to recognize similar transitions identified by others occurring elsewhere in the world. Transformations happening at Xochimilco in Mexico, Lago Atitlán in Guatemala, the valleys surrounding Cuzco in Peru, Chiloé's meadows in Chile, California's Central Valley, and innumerable other locations have contained what many of their inhabitants regard as the loss of lands possessing equally sacred and paradise-like qualities. The notion that such transformations are simply an inescapable process of modernization, however, fails to comprehend the work being conducted by a large number of agroecologists throughout the Americas. It is in this regard that agroecology offers the opportunity to pursue a very different science — along with its potential for constructing not simply a new agriculture, but a new social order as well.

This book has been written as narrative nonfiction, which means that many chapters combine extensively researched and documented material with scenes taken from the lives of people who have witnessed parts of this history.

While I have compressed this book to provide a more readable and accessible work, I am indebted to a much larger group of authors and researchers — only some of whom I have listed in footnotes. I gratefully acknowledge the contributions of many, many others.

The characters in this book are all real, having myself witnessed a large number of the scenes presented herein. In select passages, I assume limited interpretive licence and rely on my best memory of these events. Additionally, I encourage the reader to refer to others' reflections, written or otherwise, so as to better understand underrepresented perspectives of other social groups. This particular history is only the first step in appreciating a broader understanding of this world.

Note

1   *Ahupua'a* is a Hawaiian term often denoting a wedge of land extending from the top of a mountain or valley to the seashore. Ahupua'a represented a political, cultural, and social division of Hawaiian territory that recognized the interconnectedness of the land, waters, and ecology of the islands.

Part I

Conflicts in Science: Revolution and Counterrevolution in the Americas

To fully appreciate agroecology as a newly emergent science, we must examine its history. Part of this history began with a scientific mission sponsored by the Rockefeller Foundation to advance Mexico's agricultural production. This mission, established more than eighty years ago, would become a pivotal point of conflict between two scientific communities. These early conflicts fostered debates inextricably linked to Mexico's revolutionary war, with questions regarding "whose lands?," "whose production?," and "production for what purposes?" As the Rockefeller Foundation's mission spread across the Americas, another, more profound question would slowly emerge: "Whose science?" Was the science introduced to Mexico truly a neutral force as so often asserted? Or did it serve another, perhaps even counterrevolutionary purpose?

1 Scientific and Social Revolutions

"What do you mean? ... That science is political? ... How can that be?"

The questions were delivered by a distinguished professor, a celebrated authority on environmental history. I was seated at a table defending my doctoral dissertation; a study examining how a group of US scientists had dramatically altered the course of Mexican agriculture nearly one hundred years ago. The professor was about to express even greater skepticism about my study when he was interrupted by the woman sitting next to me, Cheri Lucas Jennings.[1]

"What don't you understand?" Cheri asked professor Donald Worster, while launching into her next question. "We all understand how science is supposed to be neutral, objective — supposedly without a political agenda. Yet surely you can comprehend how the group of US scientists were engaged in a political project. What is more, the political nature of their work is plainly evident in the very language, methodologies, experimental design, epistemology, and pedagogy they employed as scientists. What don't you understand about how this study reveals the political dimensions of science and scientists?"

Their argument continued for many minutes, before the faculty member leading this meeting interrupted the two academics. "Cheri, you are going to have to give your husband a chance to respond. It is after all, his responsibility to defend this work." I was greatly dismayed. As Cheri was so much more articulate about the political dimensions of science, I much preferred listening to her defence of my thesis.

This moment of academic conflict reflected a larger debate. For many people, scientific revolutions begin with scientific breakthroughs; new ways of perceiving or understanding the world around us. The observation of germs and germ theory, for example, provided a new way of understanding many diseases, rapidly changing practices in public health and medicine. It was a revolution characterized by groups of scientists coming together to gather facts, analyze the world by experimental design, and gradually convince their peers of the merits of this expanded understanding.

Scientists often portray changes in their disciplines as an internal process driven by peer review and empirical discoveries. However, a growing body of literature reveals how powerful social and economic forces can shape the direction of scientific research. Various studies have demonstrated that major shifts in scientific understanding frequently align with the interests of capital rather than emerging purely from the accumulation of knowledge.[2] This literature suggests we should view scientific communities not always as spontaneous gatherings of neutral researchers; at times they can also be deliberately assembled and oriented toward achieving specific political and economic objectives (Noble 1979; Brown 1979; Oreskes and Conway 2011).

This book takes readers on a journey into the realm of a politicized science; a journey exploring science and revolution in the Americas. At its heart lies a critical moment: the arrival of US scientists in the 1940s to Mexico with a mission to revolutionize agricultural production. Their encounter with another group of scientists supporting a contrasting mission has exposed a fundamental divide that has blossomed into a deeper and more profound conflict.

Industrial Agriculture versus Agroecology

Since the mid-twentieth century, many of the world's landscapes have been transformed into what we will describe as industrial agriculture. While its proponents claim to have saved hundreds of millions from starvation, serious doubts exist about industrial agriculture's viability as a path forward. In response, a different approach has emerged: agroecology. In the words of one of the leading voices in agroecology and his colleagues:

> Agroecology stems from the accumulation of knowledge about nature through indigenous and traditional farming and food production systems over millennia....
>
> Using their intricate local knowledge, traditional peasant farmers have maintained high levels of biodiversity associated with their farming systems, developing agroecology farming systems that cycle nutrients through closed systems, maintaining soil fertility and requiring very few external inputs. In its contemporary usage, agroecology is variously known as the science that studies and attempts to explain the functioning of agroecosystems, primarily concerned with biological, biophysical, ecological, social, cultural, economic and political mechanisms, functions, relationships and design, as a set of practices that permit farming

in a more sustainable way, without using dangerous chemicals, and as a movement that seeks to make farming more ecologically sustainable and more socially just.... Agroecology as a science and a movement has been built upon the rejection of the main tenets of the Green Revolution, namely that crop genetics and the application of purchased synthetic chemicals to monoculture is the most efficient way to produce food. The basic assumptions of high-input, industrial agriculture have led to enormous negative externalities, in terms of resource depletion, air, land and water pollution, and greenhouse gas emissions, at the same time as they have concentrated power over food systems in the hands of a few large corporations and contributed to the deterioration of the health of farm workers and consumers. Perhaps most conclusive, even on its own terms of ending hunger, the Green Revolution has failed — around 900 million people continue to suffer from malnutrition, even as half a billion now suffer from diet-induced obesity. (Rosset et al. 2020)

Even as agroecology has been inspired by the agricultural practices pursued by many millions of campesinos, it is still evolving. Indeed, agroecology offers a scientific design and practice for addressing the sustenance needs of the many while also mitigating the hazards posed by cascading environmental threats that jeopardize the very survival of human civilization in the twenty-first century.

The dominance of industrial agriculture presents several paradoxes. Even as it controls global markets, its practices have unleashed myriad social and environmental crises across the planet. This fundamental contradiction is now recognized even by many of its most celebrated beneficiaries: global agribusinesses (Rushe 2022). A related puzzle is industrial agriculture's frequent reliance on exploiting labour, drawing heavily on publicly owned and financed resources, as well as receiving massive state-supported infrastructure while failing to benefit the largest number of humans on this planet.

Given these contradictions, one might expect agroecology to have already displaced industrial agriculture in many regions. Yet it faces a hostile environment, an unexpected challenge considering the reliability and veracity of its findings. Indeed, one of agroecology's remarkable achievements in recent decades is a substantial body of scientific studies documenting its advantages over industrial agriculture (Gliessman 2016).

The mounting evidence of a worsening global crisis and the catastrophic destruction of the natural world has created an imperative to identify alternatives to the documented failures of industrial agriculture. Designing agricultural practices predicated on preserving, if not improving, the surrounding ecology would seemingly be favoured over the constancy of soil erosion, the pollution of ground and surface waters, the contamination of air basins, the destruction of habitat and species, and an array of other negative consequences so common to industrial agriculture. Yet agroecology — despite offering perhaps the most compelling alternative — confronts a variety of opposing political forces. Though it builds on traditional practices embraced by hundreds of millions of agriculturalists, it occupies a marginal place in universities, government programs, and agricultural landscapes.

A large network of scientists and researchers worldwide have documented the benefits of numerous agroecological systems. A few of these systems are vaguely familiar among visitors to Hawai'i in such features as ahupua'a and fish ponds; in Mexico's chinampas or so-called floating gardens; in terraces and waru-waru constructions in the Peruvian Andes; in Mayan polycultures in Central America, and so on (see Glossary for selected Spanish- and Hawaiian-language terms). These represent just a fraction of the systems developed over millennia and maintained by campesinos into the present.

A typical reaction among defenders of industrial agriculture is that these agroecological systems cannot match the production levels achieved by modern systems. These debates quickly spin into a world of methodological, epistemological, and related arenas. The reader will be introduced to some of these debates, but for the moment you should be aware that a focal point for many agroecologists has been to examine such questions in exacting detail. Their studies reveal that agroecological systems have high-yielding production levels, even when compared to industrial agricultural systems (Rosset 1999).

The work of many scientists in agroecology has helped to dispel a common impression of the field as merely a romanticized vision of traditional or Indigenous agriculture. Agroecologists have provided an empirical basis for documenting the features of and science behind such systems, revealing a mastery of design that blends ecology and agriculture. Another facet of their design is the low cost for maintaining many such systems (Toledo and Barrera-Bassols 2009). Pivotal for their success is a skilled and knowledgeable community, a role provided by campesinos or peasants.

The notion that agroecology is limited to traditional knowledge fixed in the past represents another misimpression of this science. Agroecological practices are being continually revised by campesinos labouring in changing environments, as well as in findings by agroecologists concerning novel practices that provide greater biodiversity. An added element of agroecology is understanding the context of historical land use practices (e.g., large estates using extensive monocropping) as a basis for understanding past barriers as well as future opportunities for creating a more biodiverse landscape and alternative marketing arrangements (Siebert and Belsky 2014).

In the twenty-first century, while vast numbers of people living and working in agriculturally based communities around the world have been described as peasants, I will use the Spanish equivalent — campesinos — when describing the mixture of people who are engaged in agricultural production on the margins of society. Such people and their communities are frequently characterized as having been left behind in the world of modern economies; a perspective implicitly arguing that campesinos need to adjust their skills to the modern world. However, from the perspective of agroecology, one gains a different view on the value of campesinos in today's world, especially when understood as a vast community possessing an increasingly invaluable knowledge about the design and maintenance of agroecological systems. As so many agroecologists in the field have demonstrated, it is often the world's campesinos who possess an intricate, practical understanding of how to blend agriculture and ecology, an expertise rarely found among those managing modern, conventional agricultural systems.

All of which takes us to a paradoxical juncture. If agroecology is so compelling and offers protection from the cascade of global crises, why has it not been at the forefront of policy initiatives, particularly those addressing climate and related crises? Moreover, if the work of agroecologists is so valuable, why do the largest number of scientists remain engaged in research and studies predicated on industrial agricultural techniques and practices? Surely if the scientific findings of agroecologists were so persuasive, would not the largest number of their colleagues across the sciences be drawn in their direction and advance such work? A further question common to so many in today's world: If agroecology were truly such a superior system to industrial agriculture, why has it not already expanded its presence around the world?

The answers to such questions are complicated by the way in which we think about industrial agriculture, progress, and the role of science. The historical reasons for neglecting the role of science in this conflict are due to the classical view that science functions as an essentially neutral force. According to this view, science-based agriculture serves humanity by demonstrating in an observable fashion its verifiable results, such as the production of food.

But what if science operates less as a framework for independent inquiry and more as a mechanism for producing specific types of knowledge? If knowledge production inherently privileges certain forms of knowing, what are the ramifications for the loss of other kinds of knowledge? More specifically, how do we understand the increasing loss of many traditional practices pursued by perhaps tens of millions of campesinos not only across the Americas but around many other parts of the world?

The history advanced in this work presents a different perspective on the role of science and progress — none of which is categorical of all science, scientific methods, or individual scientists. Rather, this history seeks to counter a myth surrounding one of the larger scientific projects of the twentieth century — the spread of scientific agriculture as advancing the interests of humanity — while providing a more detailed accounting of the political considerations embedded in its design. One theme that becomes apparent from the history presented here is the emergence of a recurring feature of scientific agriculture as practised in many parts of the world: the conflicts between the science of agroecology and those of industrial agriculture are deeply political. These issues will become more tangible in the chapters that follow; for now it is essential to explore a specific historical moment, when two very different communities of scientists would first encounter one another in the Americas.

The Rockefeller Mission in Mexico

The conflicts that now exist between these two very distinct scientific communities — industrial agriculture and agroecology — began more than a century ago with the Rockefeller Foundation and its first philanthropic ventures to rationalize society by supporting specific scientific projects. This history is instructive for understanding the Foundation's mission to advance what it would term "scientific agriculture," but before moving forward, it is useful to examine the Rockefeller Foundation and its earliest work in a variety of scientific missions.

To even mention its name — a foundation named after John D. Rockefeller — suggests a nefarious, subversive undertaking. By the beginning of the twentieth century, its founder would be known as one of the wealthiest individuals on the planet, a position gained by consolidating a nascent fossil fuel industry into what would become a globally dominant financial interest. John D. Rockefeller and his empire founded on petroleum would also grow into a global economic force, commanding more significant resources than most governments. With the rising tide of workers opposing the domination of the political sphere by a wealthy elite, the tremendous financial and political power exercised by John D. Rockefeller soon resulted in a series of laws initiated in the 1930s to break up the monopolies he controlled in oil, transportation, and their associated industries. With the success of having created a vast industrial monopoly and amassing one of the largest financial fortunes of the twentieth century, John D. Rockefeller earned a legacy as one of its most notorious capitalists. It is little wonder that the Foundation carrying his name would suggest dubious motives.

A principal reason for not focusing on Mr. Rockefeller and his role in the Foundation and the global projects it would spawn appears quite simple: its day-to-day work across the Americas was led by groups of researchers and associates sharing a commitment to science-based projects. While the Rockefeller Foundation was without a doubt consequential, the agents of change were principally a collection of scientists managing multifaceted scientific projects. In a very real sense, the divide between the vast resources of the Rockefellers and the actions of the Foundation was a cadre of scientists guided by their commitment to scientific agriculture. Yet a careful examination of this history reveals that far more was at play than the conventional view of science as a neutral, objective force in the world.

This history should not be misconstrued as an effort to demonize, impugn, or malign particular individuals or organizations. Yet neither should the reader easily submit to a belief that anyone associated with scientific work is therefore free from advancing a politically ladened project. To what extent did the Rockefeller Foundation's scientific mission in Mexico serve as an agent of counterrevolution? In later chapters we will pursue an even broader question: How do we understand agroecology as an alternative to industrial agriculture and its potential as a revolutionary science?[3]

Notes

1 The paraphrased dialogue and scenes described herein are based on the hearing of my dissertation before the Department of Political Science at the University of Hawai'i during the spring term of 1984. To be fair, Professor Worster was likely being provocative, testing whether I could defend one of the major premises of my dissertation. At the same time, Dr. Worster's questioning reflected a common perspective among his peers across many disciplines — that science and technology are largely neutral and without political valence. Dr. Worster's many publications include *Nature's Economy* (1977), *Dust Bowl* (1979) and *Rivers of Empire* (1985).

2 The differences between Ludwik Fleck (1981) and Thomas S. Kuhn (1970) on science and scientists represent part of the philosophical/epistemological bedrock of examining the structure of scientific revolutions in this work. It is especially Fleck's earlier work in 1935, which pointed to the social dimensions of scientists' thought collectives and the role of popular participation as a guard against the "harmony of illusions" that might otherwise prevail among groups of scientists, that presents a central question regarding the relationship between science and politics.

3   The reader should remain mindful of the rapidly changing political terrain in which scientific findings, methods, and facts are being increasingly dismissed without substantial evidence (e.g., President Trump and his allies' rejection of climate change as a hoax and of vaccines as hazardous). Nothing in the arguments presented herein should be interpreted as supporting such categorical rejections of scientific standards and methods. The history presented in this book is focused instead on the manipulation of science and scientists to achieve certain social, political, and economic objectives. It is particularly the limitations placed on science projects — such as excluding a more rigorous analysis of the consequences of scientific and technological projects as well as a broader comparison of different food systems — that raise questions about the political nature of the sciences and scientists supporting industrial agriculture versus agroecology and agroecologists. Nothing in this work is intended to demonize any scientist. This history is intended to inform readers about the political nature of science, especially as a means of comprehending competing scientific communities, their relationships to money and politics, and the resultant consequences in the world.

2 Transforming Mexican Agriculture

One of the most consequential efforts by private philanthropy to reshape agriculture began with three American scientists crossing the Mexican border in the summer of 1941. The Rockefeller Foundation's official narrative presents this moment as an apolitical scientific mission — merely an attempt to help Mexico solve its pressing agricultural problems, such as the absence of infrastructure applying modern science and technology to increase agricultural production. Yet this seemingly straightforward story of technical assistance masks a more complex history — one that raises two fundamental questions: What were the pressing problems in Mexican agriculture? And what measures did the team of US agricultural scientists take to address them? It is worth considering the version published by the Rockefeller Foundation:

> In Syracuse [New York], Bradfield picked up Paul Mangelsdorf, a botanist from Harvard University who was one of the world's leading authorities on corn. Mangelsdorf brought with him a young Harvard graduate student, Richard Schultes, who was an adventurous scientist — part botanist and part anthropologist ... his greatest contribution was his ability to speak Spanish ... Dr. Elvin Charles Stakman, a plant pathologist from the University of Minnesota, completed the team. Stakman was the most prestigious scientist on the trip: older, and first among equals. (Among his graduate students at Minnesota was a young plant pathologist by the name of Norman Borlaug, whose interest in wheat stem rust would later make him a hero of the Green Revolution.) .... These were not junior researchers. They were the foremost scientists in their fields. It was testament to the authority of the Rockefeller Foundation that when President Raymond Fosdick asked who might serve on a commission to study how the Foundation could help Mexico, three of the most prestigious agricultural scientists in the nation answered the call and gave up their summers to chase the opportunity.

Heading south toward the Laredo border crossing, the station wagon rumbled through cotton farmers in east Texas where the General Education Board (GEB), another Rockefeller philanthropy, had first experimented with agriculture programs 30 years earlier. The GEB had been a primary sponsor of the farm demonstration movement that agricultural scientist Seaman Knapp led at the turn of the century. Knapp had taken a holistic approach to agriculture, combining science with popular, demonstration-based education for farmers. His work had long since been popularized and taken over by the extension program at the US Department of Agriculture (USDA) and the nation's land grant colleges. It was just the way the Foundation worked. Invest early in an innovative idea, bring a program to maturity, and then pass it to a government or other entities for permanent support....

History may record that Stakman, Bradfield, Manglesdorf and Shultes were groundbreaking pioneers, but as they drove through Mexico, they traveled in the footsteps of others. For almost a decade prior to their trip, Dr. John Ferrell, associate director of the International Health Division of the Foundation, and Josephus Daniels, US Ambassador to Mexico, had tried to convince the Rockefeller Foundation to pursue agriculture work in Mexico. They saw it as an opportunity to work on a pilot project to determine if elevating the farm economy could improve nutrition for the rural poor. Their early attempts to convince the Foundation failed. It would take a shift in the Mexican political landscape, the spread of war, and the addition of US government voices to the chorus of advocates for Mexican agriculture work to convince the Foundation to explore it as a possibility. With all of these elements aligned, in early 1941, Foundation President Raymond Fosdick agreed to convene a study commission to collect data. What was the situation in Mexico? How could the Rockefeller Foundation help? A few months later, Bradfield and his colleagues crossed the border in their green carry-all station wagon....

[They] were simply a scientific survey team with a mandate from the Foundation to explore ways to help Mexico solve its pressing agricultural crisis. Truth be told, they were not exactly sure what could be done. There was no precedent, no template for what they were about to do. They intended to drive into Mexico and take it one mile at a time. (Rockefeller Foundation 2013: 19–23)

So begins the Rockefeller Foundation narrative describing what would eventually become a network of international organizations transforming agriculture on a global scale. On its face, sending a team of agricultural scientists from the US to Mexico to investigate the source of problems in agricultural production and possible solutions seemed like an unproblematic and apolitical way to serve humanity.

Decades later, the former secretary-general of the United Nations, Kofi Annan, would celebrate the Foundation's work in Mexico:

> The Foundation launched a remarkable initiative in Mexico to increase food production substantially through the development and introduction of more productive and resilient varieties of wheat, corn, beans, and other staple crops. The increases in agricultural yields were spectacular. Over the next several decades the Rockefeller Foundation sought to introduce high-yield seeds and new cultivation strategies in other developing nations. These efforts became known as the Green Revolution and are credited with saving more than a billion lives.
>
> Certainly the Foundation and the world learned lessons along the path of the Green Revolution. In some regions, greater agricultural productivity heightened inequality and intensified the marginalization of the poor and vulnerable in society. In other places, intensive use of petrochemical fertilizers and irrigation led to environmental problems and even the increase of human parasites like schistosomiasis. These consequences prompted serious reflection and great debate both within and outside the Foundation ... the Rockefeller Foundation learned from success and disappointment. Deepening its commitment to fighting hunger and malnutrition after the 1970s, it continued to invest in both science and human capacity. A second phase of the Green Revolution focused on continued increases in agricultural yields, while striving to protect the environment and strengthen communities along the way. (Rockefeller Foundation 2013: 15)

Like other such summaries, the language of "mistakes were made, lessons were learned" disguises a more complex and conflicted history. Were these "mistakes" in contemporary agriculture not intentionally incorporated as an integral part of the Rockefeller Foundation's design of "scientific agriculture"? In the case of Mexico, the Foundation's scientists operated as a decisive force in defining a particular approach to advancing agriculture and society. A story about three guys in a car having only a scant idea about what they might propose was not a convincing narrative for a plan that had been years in the making. Similarly, explanations that the Rockefeller scientists were simply working to provide more food, as the reader will learn, obscured a more complex terrain of politically laden choices (e.g., what foods were being grown for whom at what prices and with what consequences?). Far from being an unproblematic application of science and associated technologies, the mission in Mexico reflected decades of experience managing teams of scientists for purposes beyond merely delivering the best science.

When the Mexican government and the Rockefeller Foundation formalized their agreement in 1943, they confronted fundamental questions about the nature and purpose of agricultural knowledge. Often characterized simply as a technical assistance program to establish "scientific agriculture," the Foundation's mission in fact operated at the intersection of competing visions for Mexico's agricultural future. These conflicts extended far beyond technical matters to encompass crucial decisions about which farmers to support, what political alliances to form, how to structure market relationships, and whose expertise to value.

Over the next two decades, the mission would prove to be a decisive instrument for transforming broad facets of Mexican society, economy, relationships between public and private spheres, as well as science and nature. The program focused on increased agricultural yields per hectare as an essential mechanism for feeding the world. Yet even in its early years, it faced a number of junctures in its journey where controversies arose regarding the direction and purpose of its work in Mexico. Understanding the paths *not* taken — particularly in terms of addressing adverse consequences — is as important as the record of its claimed achievements.

Resolving debates about alternative paths would not and could not be achieved by resorting to a neutral scientific method. In this sense, one of the more profound elements of the Foundation's work in Mexico centred not merely on which disciplines to empower; equally important were decisions about what kinds of knowledge to dismiss as having little or no value. The Foundation's management of science engaged in a fundamentally political act throughout its work in Mexico, an endeavour referred to here as the production of knowledge.

Other facets of this mission that moved its work well beyond any common meaning of a science-based project were the overt political activities of its scientists. Most striking were decisions about where to locate its fieldwork — where to organize field demonstrations, working with which groups of growers, which crops to select, as well as what metrics to apply for evaluating particular kinds of consequences. These considerations emerged as points of conflict not easily resolved by resorting to a scientific rationale. Indeed, one of the earliest decisions to be made was precisely who would be the first and foremost beneficiaries of its work.

This question of beneficiaries stood at the heart of a deeper institutional conflict. Mexico's post-revolutionary 1917 constitutional and legal framework had established clear priorities: agricultural development should primarily benefit the millions of impoverished campesinos working small plots of land. The background to the favouring of campesinos was that until 1910 Mexico's agriculture was commanded by approximately six thousand large estates, haciendas, that occupied 94 percent of Mexico's total land surface — whereas perhaps more than ten million campesinos held less than 1 percent of the land (Toledo and Barrera-Bassols 2017). In the early 1940s a number of the most prominent officials within Mexico's Ministry of Agriculture advanced policies supporting collective over private property. Among the different branches of the ministry, advocates for campesinos occupied positions of authority most prominently in the Institute for Agricultural Investigations, or the IIA as it was known by its Spanish acronym. This group of researchers, drawing on popular themes of Mexico's revolutionary period, viewed their work as serving the vast majority of Mexico's agriculturalists — its millions of peasants.

Such a perspective, however, was not broadly shared among the Foundation's scientists. In order to achieve its aim of high productivity, it became necessary for the Foundation to transform institutional relationships to meet the new requirements of scientific agriculture. The Foundation thus established a separate office within the ministry, the Office of Special Studies or OEE, which would be directed by Foundation-affiliated scientists. The rationale for establishing the OEE as a separate office at the inception of the Rockefeller mission reflected a clear understanding of basic conflicts with others in the ministry, particularly the IIA, regarding the solution to problems in Mexican agriculture. One scholar of Mexican agriculture characterized the officials at the IIA as fundamentally opposed to the OEE. While the IIA shared the objective of Rockefeller scientists to achieve increased production, the concept of whom the science should serve differed radically: "They [the IIA] believe that increased production in the Mexican countryside was intimately linked to structural changes which would break up large capitalist farm holdings and create viable cooperatives of peasants and workers in their place" (Hewitt de Alcántara 1976: 192).

How did this view oppose that of the Rockefeller Foundation? What path would the mission take to achieve higher production, and what portion of society would that path come to serve? To more fully comprehend the political nature of the Rockefeller Foundation's support for certain sciences and whom it would serve, it is helpful to examine how they came to engage in "the management of science."

A Science for Whom?

At the beginning of the twentieth century, the Rockefeller Foundation began what would become a multifaceted and decades-long effort to "improve" society and social relations in the United States. Following the creation of the Rockefeller Institute in 1901, John D. Rockefeller approved the launching of a more targeted effort by establishing the General Education Board (GEB) in 1903. During the early 1900s the GEB, among its varied efforts, focused on instructing farmers about modern scientific agricultural techniques. One of the architects of the Foundation's program expressed the value of such an approach to achieve larger purposes: "Promotion of the development of science in a country is germinal; it affects the entire system of education and carries with it the remaking of civilization" (Anonymous 1950: 1).

The first place that the GEB concentrated its work was in the southern United States, placing special emphasis on what it termed "scientific agriculture" (Bradfield et al. 1951: 192). At a basic level, this referred to a practical approach that involved disseminating technical information about when, where, and how to plant a particular crop.

In the following eight years (1906–1914), the Foundation would expend nearly $1 million on its program in scientific agriculture, resulting in excess of 100,000 demonstrations of techniques designed to make agriculture more productive. For the Foundation's officers, the management of this technical knowledge via such demonstration rested on a more essential social and political act: to generate support for these programs by reinforcing its value to merchants, bankers, wealthy farmers, and the nascent beginnings of agribusinesses. These agricultural demonstration projects served to not simply deploy new technologies but "as a means for changing the social order" (Cleaver 1975: 146–7). The Foundation's strategy, or "scientific agriculture," was premised on building a new political economy in the US South:

> Convince the owners of farms who reside in town that there is a way to get more rent; drive home the thought to the merchants that low earning capacity limits purchasing power, circumscribes trade and casts the constant shadow of uncertainty upon the day of settlement; awaken the banker to the fact that it is unwise to loan to men who farm the best land on a fourth of a possible

crop, and poor lands on a tenth ... convince and arouse this land proprietor, this merchant and this banker, and they will not only give their influence, but will insist that all their tenants adopt the new method. (Ferrell 1936)

But the Foundation's perspective on transforming the social order was not simply limited to the science of agriculture, nor was its vision for advancing healthy economies limited to just the United States. In 1913 the Rockefeller Foundation established a public health mission in Mexico that would prove instrumental in building institutional relationships. By the early 1930s, this mission had trained numerous Mexican students in public health at US universities, many of whom returned to occupy strategic positions in Mexico's national health department (Ferrell 1936). Dr. J.A. Ferrell, a Rockefeller Foundation officer involved in public health, expressed concerns about Mexico's economic development; he reported that one of the country's most urgent needs was a program to improve economic conditions. However, because the average Mexican family was too poor to pay taxes, Ferrell stated that funds were unavailable to finance such services in health, education, and welfare. The solution, in his view, lay in fundamentally restructuring Mexico's agricultural economy.

In 1936 Dr. Ferrell spoke with a former Mexican minister of agriculture who had served under President Plutarco Calles about the possibility of establishing a cooperative mission in agriculture between the Rockefeller Foundation and the government of Mexico. The former minister of agriculture assured Dr. Ferrell that such a mission would be welcomed by the current president, Lázaro Cárdenas. Ferrell then wrote a memo to the president of the Rockefeller Foundation about establishing this agricultural mission: "A beginning toward economic improvement could be undertaken in the field of agriculture. No serious controversial issues would be involved and ultimately the range of aid might be broadened ... I believe two or three qualified persons might be sent to study its agricultural problems and possibilities and then outline a broadly constructive program" (Ferrel 1936: 1).

The reasons surrounding the Foundation's inaction on Dr. Ferrell's proposals for a period of years are not documented; nevertheless, it is not unreasonable to assume that one of the explanations for the pause were the revolutionary policies pursued by President Cárdenas. By 1936, Cárdenas had accelerated the break-up of latifundios (historically large landholdings) and assigned these to campesinos, many of these organized into collective ejidos or communal landholdings. As will be shown in later chapters, this vision of agriculture — based on communal ownership and campesino farming — ran counter to the Foundation's model of market-driven transformation. The Foundation's experience in the US South, which relied on bankers, businesses, and government agents compelling farmers to adopt new methods, could not easily be applied to Mexico's campesinos, who lacked both capital and connections to financial networks.

With the election of a new Mexican president, Manuel Ávila Camacho, in 1940, the pause in the Foundation's action was lifted. By early 1941, the Foundation's officers agreed to organize and send a survey team as suggested in Dr. Ferrell's 1936 proposal. The "qualified persons" selected to review Mexico's agriculture included three agricultural scientists: Dr. E.C. Stakman, professor of plant protection at the University of Minnesota; Dr. Richard Bradfield, professor of soils and agronomy at Cornell University; and Dr. Paul Manglesdorf, professor of plant genetics and breeding at Harvard University (Bradfield et al. 1941).

As detailed in the Foundation's narrative, in July 1941 the three American scientists travelled to Mexico. By December the team had completed its survey and issued an evaluation of the feasibility for launching a technical assistance program based principally on five elements:

1. Breeding improved varieties of maize, wheat, and beans;
2. Developing improved agronomic production-management practices;
3. Improving weed control;
4. Improving animal production;
5. Training a corps of Mexican scientists.

In a confidential memo to the Rockefeller Foundation, the three scientists provided a straightforward strategy — one that did not begin with the peasantry but "started at the top":

> The plan presented assumes that the most rapid progress can be made by starting at the top and expanding downward. The alternative would be to start at the bottom and work toward the top. A program of improving vocational schools of agriculture and extension work directed toward farmers themselves might be undertaken. But the schools can hardly be improved until extension men are improved; and investigational work cannot be made more productive until investigators acquire greater competence. (Bradfield et al. 1941)

While their recommended strategy reflected the institutional and political model operating in the United States, it begged answers to a number of obvious questions.

Were the most important problems in Mexican agriculture technical in nature? Even assuming that plant breeding, plant protection, and improved agronomic practices were foremost for consideration, whose agriculture was the focus for these improvements? And perhaps most concerning was the entire question of competency. One of the seemingly unarticulated premises advanced by the scientific team was that those possessing advanced scientific training were more competent for addressing what was broken and needed to be fixed in Mexico's agriculture.

Questioning the assessment of Mexico's agriculture by a select group of US agricultural scientists might have been consigned to the netherworld of empty critiques ... except for the fact that the Foundation received just such a critique as the program of technical assistance was becoming formalized and launched as one of its major philanthropic efforts. Prior to the survey team's visit to Mexico, an officer of the Foundation suggested that the views of another person be solicited regarding the proposed technical assistance program, someone familiar with Latin America. Dr. Carl Sauer, a professor of geography with the University of California at Berkeley, was suggested as a person who was "in entire sympathy with the importance of focusing on agriculture" (Bradfield et al. 1941). The Foundation was acquainted with Sauer through his proposals to the Foundation and his work on projects in Latin America.

In his memo to the Foundation, Dr. Sauer recommended that the improvement of the genetic base of agricultural crops be predicated on an understanding of the poorer segments of society. As an example, Sauer suggested that improvements in agricultural production could be placed in the context of the role of various foods among poor households. The nutritional practices of the Mexican peasantry or campesinos, Sauer stated, were excellent as far as their pocketbooks allowed. The same was true of their agricultural practices. Sauer emphasized that the main problem confronting Mexican households was one of economics, not culture (Sauer n.d.).

In concluding his remarks on agriculture, Sauer reminded the Foundation's officers that plants such as maize had much more varied use in Mexico than was true of the same plants in the United States. As a result of these differences, Sauer cautioned against applying the agricultural sciences to recreate the history of US commercial agriculture in Mexico:

> A good aggressive bunch of American agronomists and plant breeders could ruin the native resources for good and all by pushing their American commercial stock. The little agricultural work that has been performed by the experiment stations people

here has been making that very mistake, by introducing US forms instead of working on the selection of ecologically adjusted native items. The possibilities of the disastrous destruction of local genes are great unless the right people take hold of such work. Additionally, Mexican agriculture cannot be pointed toward standardization on a few commercial types without hopelessly upsetting the native economy and culture. The example of Iowa is about the most dangerous of all for Mexico. Unless the Americans understand that they'd better keep out of this country entirely. That must be approached from an appreciation of native economies as being basically sound. (Sauer n.d.: 1)

Even though Dr. Sauer's memo was distributed to several members of the Foundation — including John Ferrell — and would generate discussions among its officers extending over a period of years, his fundamental critique of a technical assistance mission patterned after the model of Iowa would remain largely unexamined for decades to come.

Among the questions Sauer raised about the Foundation's mission were the strategies of "starting at the top," or what others within the foundation world would later refer to as "betting on the best." In this regard, Sauer forwarded another memo to the Foundation in February 1945 based on a meeting with a scientist who would eventually direct the mission's program on maize in Mexico, Dr. Edwin Wellhausen. In commending Wellhausen's work, Sauer wrote, "he sees that they (the other US scientists working in the Mexico mission) must work with the native corns; he is not a missionary for soybeans, and I suspect he sees the pitfalls in the wheat campaign" (Sauer 1945: 1). Sauer continued his memo, placing particular criticisms on the Foundation's expanding efforts to increase wheat production in other countries. Sauer argued that local varieties of corn were of much greater value to local communities than would be the research pursued by the MAP (Mexican Agricultural Program, a term the Foundation used when referring to its work in Mexico) on barley, alfalfa, and wheat. He warned that these other grains had little place in many local economies, and they were ill-suited not only to the economy and ecology of Mexican villages, but in Peru and Chile as well (Sauer 1945: 1).

Complementing Sauer's observations was another set of studies undermining the Foundation's program in Mexico — studies produced by one of the Foundation's long-standing programs on examining the nexus between health and agriculture. Between 1942 and 1945 the Foundation's International Health Division conducted a series of nutritional studies in cooperation with the Mexican ministry of health. These village-based studies represented one of the best windows on understanding the relationship between agricultural practices and diet. The conclusions from one in the series of studies (examining campesino households in the state of Mexico) cast doubt on changing traditional diets: "Attempts to change dietary habits in this region would be a mistake until economic and social conditions can be improved" (Anonymous 1944: 34). It was a conclusion mirroring what Sauer had written only a few years earlier, stating that native people of Mexico "need to be encouraged that their ways are good and they need protection against exploitation" (Sauer n.d.: 3).

All of this led Sauer to voice once again his criticism of deploying the United States' version of agriculture as a model of development: "The agricultural situation in Mexico is an exaggeration of that which we have in our experimental work in California; the interest is directed away from subsistence or village agriculture to the needs of city or factory with the attendant emphasis on standardization of product and on yield, also on tariff protection, and on the commodities which the privileged fraction of the population can absorb" (Sauer n.d.).

Professor Sauer's commentary highlighted the notion of agriculture as serving society along class lines: standardized cheap foods for an urban proletariat, the development of commodities disconnected from the needs of rural economies, and the provision of luxury foods for the bourgeoisie. This portrayal provides a haunting image of what many observers decades later would decry as a fundamentally flawed structure of agriculture.

In the years to come, Sauer's cautious observations would underscore the negative consequences inherent in a technical mission dedicated to replicating the model of Iowa — a model defined by expanding the reach of unrestrained markets dominated by private interests.

Among the first in a line of cautionary statements was a memo advising the Rockefeller Foundation's president to end the Mexican Agricultural Program in the 1950s before its work "presents problems not now even dimly perceived by many Mexicans" (Dickey 1951: 3).

3 The Counterrevolution in Mexican Agriculture

No one who witnessed the original Rockefeller Foundation mission is alive today. While certain members of the original Foundation team of scientists left behind their accounts of the time, gaining the perspective of the Mexican participants from written records is scant. There is, however, a striking visual representation of the time that was no doubt experienced and shared by both a large number of Rockefeller scientists but especially an even larger group of young Mexican students and faculty who worked among the US team.

A powerful visual record of Mexico's revolutionary ideals survives from the late 1930s in the vast murals painted by Diego Rivera and contained in a chapel on the grounds of a prominent agricultural college in Mexico, Chapingo — a campus that served as a de facto headquarters for the Rockefeller mission during its early years. Rivera's murals at Chapingo point to a variety of themes, some of which, such as the role of technology, are complicated. Yet most of the chapel's panels convey an unmistakable message: the heroes of the recently fought Mexican revolution are the masses of campesinos who have thrown off the shackles of an oppressive elite.

If one were uncertain about Rivera's message in Chapingo, the same theme is expressed in an even more dramatic fashion in one of his largest works at the Presidential Palace in Mexico City. These murals, while celebrating an even older and broader swath of Indigenous peoples rebelling against the rule of foreign invaders, unmistakably condemn the wealthy European elite as a corrupting and malign force in Mexican history in visual format.

If one were to arrive in Mexico during the 1940s from the United States and witness these artistic statements portraying the forces for good and evil in Mexican politics, two very strong and popular sentiments would become apparent. First and foremost, assisting the heroes of Mexico's Revolution — the campesinos — was a good thing. Particularly at Chapingo, students and faculty would consistently receive the message that their purpose was to advance an agricultural system where campesinos were the primary beneficiaries. Second, if anyone associated with the Rockefeller Foundation was to be compared to particular images in Rivera's murals, it was clearly as members of a foreign elite.

It was little wonder the Rockefeller Foundation hesitated for several years to initiate a program in Mexico. It also helps to explain why the Foundation might have had a strong motive to present its work as anchored in a more politically neutral frame. Indeed, even for Rivera's murals at Chapingo, the role of science and technology appears in a politically neutral fashion.

Decades later, those changes wrought in Mexican agriculture might appear as simply the process of modernization, a process that unfolded across much of the world following the end of World War II. From this perspective, the Rockefeller Foundation's mission in Mexican agriculture might be understood as simply an early engine of economic development. Indeed, many histories of this period frame this and other such programs of technical assistance as grounded in the advancements and applications of scientific knowledge. Questions about whose knowledge and whose science are not discussions that come easily to mind.

A closer examination of the Rockefeller Foundation's mission, however, reveals a science with a deeply embedded political agenda. More than simply a project predicated on sharing scientific advancements, the Foundation's early work emphasized the role of a "managed science," reflecting an interest with what its officers often referred to as "rationalizing" society. In the early 1940s, it was immediately evident that this view often drew its representatives into precarious political terrain. Determining who held the mantle of scientific authority as well as who did not at various junctures threatened to reveal the political nature of the organization's work.

One of the Foundation's prime directives was to "start at the top" and "bet on the best" as a method for determining where to initiate its technical transfer of scientific agriculture. In practical terms, this meant so-called experiments would be based on experimental conditions that favoured rapid increases in crop production. It was at precisely this juncture where the Foundation's scientists encountered a conflicting perspective about their experimental design.

A group of Mexican scientists working at the IIA thought that Mexico had the opportunity to pursue a very different path: starting at the bottom. They countered that by focusing on those campesinos working in the most challenging conditions, the mass of Mexico's poor campesinos and their communities could benefit from raising production among those most in need. This contrasted with the focus of Rockefeller scientists working at the OEE, who advocated working with those farms possessing access to fertilizers, pesticides, machinery, and capital to achieve a tremendous increase in production per unit of land.

In many respects, the contrary missions of the IIA and OEE might be seen as simply different scientific experiments. From the perspective of the Rockefeller scientists, betting on the best simply reflected what they knew and had experienced in places such as Iowa or California. Concentrating agriculture and capital among farmers who already possessed the wherewithal to achieve huge increases in production, at least in its early years, was regarded as a successful demonstration of scientific agriculture. For the US scientists, the central, broken feature of Mexico's agriculture was trying to spread limited resources among too many campesinos.

The resolution on which mission to pursue was complicated by a fundamental question: if increased food production for the largest number of Mexicans could be achieved by a strategy focusing on campesinos — beginning at the bottom — did it not make more sense to begin a program of assistance that would direct resources to them? For the IIA scientists, this reasoning was compelling. If even a smaller increase in yields could be achieved, but through a much larger base of campesinos working on millions of hectares, would the IIA approach not be superior to the Rockefeller team's approach, beginning with a vastly smaller number of farms, even if these larger holdings achieved tremendous increases in production? As later summarized, the IIA approach in the early 1940s should have given pause against the Rockefeller model:

> Since irrigated corn accounts for only 5.3 percent of the area sown annually, if yields are increased 50 percent only in irrigated regions, total production will only increase 5 percent. The best possibility for raising output is to be found in a modest increase in the yield per hectare [of corn] in the non-irrigated regions which constitute 94.7 percent of the total sown area. (Hewitt de Alcántara 1976: 39)

There could be many reasons for favouring one experimental design over the other, but one thing was certain from the start: the modest increase in production for poorer campesinos that IIA scientists advocated would entail a different project in terms of likely social outcomes. On the face of it, there was no compelling scientific rationale for pursuing one design over the other. The issue was political: which was more worthy of prioritizing? It was precisely at this juncture that the Foundation exercised its own judgment about which research held greater value.

The Foundation would continue to confront arguments about the nature of its mission in Mexico in the coming decades. In place of a healthy debate, the Foundation's scientists and their managers demonstrated an unwillingness to consider an alternative mission. To illustrate this lesser known history, let us return to the discussions occurring inside the Rockefeller Foundation before it had completed its first decade of work.

Even though the MAP had made important strides with demonstrating the Iowa model in a few short years, the Foundation's officers were aware of controversies suggesting it may be necessary to conclude its work in Mexico. In 1949 Warren Weaver, then president of the Rockefeller Foundation, invited three of the Foundation's trustees to Mexico to witness the mission's achievements. Following their visit Weaver asked the trustees to consider two questions: (1) What are your ideas about the method for concluding our work in Mexico?; and (2) How can we capitalize on the useful administrative, organizational, and technical procedures developed by the mission?

The response provided by one trustee, Dr. John S. Dickey, argued for the early conclusion of the Foundation's work in Mexican agriculture; he suggested that a "limiting principle" should be solely based on its scientific utility:

> I suggest that one of our limiting principles might be that we will not carry the program beyond the point where it has been established as a scientific success; to put it otherwise, that the Foundation is not and will not be primarily concerned with the long-range problems of practicality involving major political, economic, and social decisions and operations. (Dickey 1951: 2)

Dickey's advice to the Rockefeller Foundation was based on his own economic and political consultancy. Beginning in 1940, Dickey was appointed a special assistant to the US secretary of state; he was responsible for establishing US trade and commercial policy throughout the Americas. It was also in this position where he worked directly with Nelson A. Rockefeller, the coordinator for inter-American affairs and son of J.D. Rockefeller. His blunt rationale revealed a stark political calculus for distancing the Foundation's technical and scientific work from its social consequences:

> For example, I can imagine that this program before long might begin to have a considerable impact upon the whole land-use policies of Mexico, and I am perfectly sure that within three to five years the program will raise some very acute problems with respect to the political control of these benefits ... these very

benefits may introduce fresh economic disparities within the Mexican economy, which will present political problems not now even dimly perceived by many Mexicans. (Dickey 1951: 2–3)

Dr. Dickey concluded with a bleak forecast about the trajectory of the Foundation's work within the next decade: "It would be unfortunate for all concerned, especially for the program itself, if the Foundation is heavily in the picture when this [growth in social tensions] takes place" (Dickey 1951: 2–3).

Dickey's counsel begged more profound questions surrounding the dividing line between science and politics in Mexico's agriculture. On its face, Dickey's comments were much less instruction about how to maintain an authentic program of scientific inquiry and much more a document for pursuing a longer term political strategy.

At the intersection of this conflict — advancing a seemingly neutral science-based program, yet one likely to generate negative impact affecting millions of peasants — stood the Rockefeller Foundation, its scientists and their managers as arbiters about how to address this apparent contradiction. Dickey's answer — to create distance between the advancement of a science-based agriculture and its negative consequences — reflected a narrative the Foundation had already practised for many years and would continue into the future. The solution, of course, was to pursue an agriculture based on the conditions faced by millions of campesino households. What is more, the Foundation already possessed a contemporary critique paralleling Dickey's assessment, but with a radically different conclusion: the Foundation needed to avoid pushing the model of Iowa.

However, John Dickey's concerns were not predicated on how to construct a science-based program that would benefit the vast number of Mexicans involved in agriculture. Rather, his anticipation of the negative consequences sure to follow from the Foundation's work was to suggest a strategy whereby the Foundation could avoid its responsibility for having unleashed a host of negative consequences. Instead of measuring the Foundation's experiment in agriculture in terms of improving the social conditions for millions of peasant farmers, the program in Mexico would be constructed around a set of metrics having value in terms of the accumulation of profits by a specific class of farmers, linkage to larger agribusinesses, and its integration with larger financial interests operating regionally, nationally and internationally.

It is worth pausing to consider the contrast between Dickey's memorandum and later narratives about the Green Revolution explaining that "mistakes happened, but lessons were learned." In Dickey we have direct evidence of a warning to the highest levels of the Foundation that bad things were surely soon to follow as a result of their work in Mexican agriculture. A warning not unlike that presented by Carl Sauer and dismissed by the Foundation years before. By 1951, not only is Dickey's warning not dismissed, it argues that things are breaking and the Foundation should prepare for its departure before the link to its responsibility for this destruction becomes more apparent. The lesson, seemingly for Dickey, was that the Foundation should distance its "scientific work" from the negative social consequences following from its achievements, which only increased production and profits for select farmers, agribusinesses, and financial interests while the lives of many campesinos worsened.

The common narrative that the program be referred to as a cooperative scientific and technical assistance endeavour misrepresents the nature of its activities in Mexico; the Foundation's confidential record of the time point to a more specific set of social, political, and economic objectives. In 1950, the archived memo from former Foundation president Warren Weaver to acting president Chester Barnard illuminated the architecture of a program that would never again appear in the public narratives contained in the Foundation's explanations of its mission in Mexican agriculture:

> We customarily refer to this program as a collaboration between the Mexican Government and the Rockefeller Foundation, this collaboration having been undertaken at the request of the Mexican Government. Although formally accurate, this statement is actually very misleading…. It must be realistically admitted that they had little or no idea as to what we were talking about or what we intended to do. (Weaver 1950: 1)

Weaver went on to explain that the Foundation provided little opportunity for Mexicans to share in decisions as the joint program progressed:

> First of all, if one takes into account the fact that this program was initially more or less stuffed down their throats, and that all of the money involved is spent under the complete control of North Americans [a flawed reference to United States], then it must be conceded that Mexico has in fact increased their contribution, year by year, in a rather surprising way. (Weaver 1950: 1)

At the time Weaver completed this memo to his colleagues, the Foundation had only begun to launch its work in Mexico. Its most important achievements — ones that would provide a template for many of the global research centres that would follow — were yet to be achieved.

It would become apparent that the principal flaw in Diego Rivera's portrayals of the political forces at work in Mexico was misunderstanding science and technology as simply neutral forces in the world. As the Foundation's internal documents demonstrate, scientific expertise served as a powerful tool for advancing a counterrevolutionary agenda in Mexican agriculture. By framing their work as purely technical assistance, the Foundation obscured how their chosen approach to agricultural development would systematically undermine the revolutionary gains of Mexico's campesinos. The question facing the Foundation at this juncture was how to institutionalize a political agenda that would endure for many years to come.

The Production of Knowledge

At the inception of the Mexican Agricultural Program, it was necessary to establish a headquarters for its mission. The obvious choice was the Escuela Nacional de Agricultura at Chapingo, not far from Mexico City. Chapingo offered expansive lands where Foundation scientists could demonstrate their superior production methods to students, farmers, and policymakers. The fact that this major university had emerged from the seizure of a former hacienda controlled by one of Mexico's pre-revolutionary elite did not appear to be consequential to the Rockefeller Foundation. Nor did the fact that in 1924 Chapingo's post-revolutionary leaders expressed the school's purpose as advancing social justice for Mexico's campesinos, along with death to the latifundistas and hacendados. One would think this might have served as a more cautionary message to the Foundation (Caire-Perez 2016). However, within a few short years, many of the Foundation's agricultural scientists would serve as both faculty at Chapingo as well as architects for training subsequent generations of students.[1]

Accompanying its program to increase the production of select commodities, the Rockefeller Foundation also aimed to produce professionals who would guide Mexico's agricultural development in the coming decades. By the end of the 1950s the Foundation supported advanced training for hundreds of young Mexicans, with advanced studies at land grant universities in the United States. This dimension of the Foundation's work — the production of knowledge — would constitute a powerful instrument for transforming agriculture in Mexico and eventually many other parts of the world.

The Rockefeller Foundation's training programs not only shaped agricultural science but also had profound political implications. These dynamics are best understood through the experiences of one trainee who would become both a scientist and a challenger of the Foundation's agenda: Efraím Hernández Xolocotzi.[2] Xolocotzi would eventually be recognized for his extensive studies of botany and ecology throughout the Americas, earning degrees at Cornell University and Harvard University, and for his work with both the Office of Special Studies (1945–1949) and its successor organization, the International Maize and Wheat Improvement Center (1968–1972), or CIMMYT as it was known by its Spanish acronym. By 1955, Xolocotzi would begin a teaching career at Chapingo's National School of Agriculture.

In addition to being schooled in the United States, Xolocotzi was mentored by one of the three founding scientists who advised the Foundation to establish the mission in Mexico. Yet there were elements in Xolocotzi's orientation as a scientist that set him apart from the Foundation's senior scientists. It was especially his training as a botanist/ecologist intrigued with Mexico's living landscapes that distinguished Xolocotzi's work from the typical projects conducted by the Foundation's other scientists.

Xolocotzi began his studies in the countryside, learning about Mexico's agriculture by listening to local peasants and gaining from their knowledge of plants, their varied uses, and the surrounding ecology. One of his earliest proposals emerged after he learned about a problem for peasant households — one stemming not so much in field ecology but from economic conditions. The specific threat to the livelihoods of peasant farmers in Durango surfaced with the crash of international commodity prices for cotton during the 1950s.

Xolocotzi's proposal to the program offered insight into a very different science-based project than those typically pursued by the Rockefeller mission, and it held the potential to place the Foundation's mission on a very different path.

> By the 1950s Hernandez had made a reputation for being willing to go against the grain. The OEE [Oficina de Estudios Especiales — the Spanish acronym for the Office of Special Studies] at one point employed him to study alternative crops in Durango. With the fall of cotton prices in the mid-1950s, Xolocotzi proposed that the OEE could assist local peasants by substituting a crop that was salt and drought resistant in conjunction with introducing goats for breeding ... thereby helping peasants to maintain a basic income. Supervisors for the project terminated Xolocotzi's participation after rejecting his proposal. (Caire-Perez 2016: 107)

While the details surrounding the rejection of Xolocotzi's proposal are not fully known, his project was clearly at odds with the Foundation's dominant approach of working with Mexico's elite farmers. Xolocotzi's proposal to address the obstacles facing small landholders on marginal lands with little access to credit or natural resources (e.g., water) was unlike any of the Foundation's usual agricultural research. Indeed, a proposal addressing the economic plight of campesinos represented an area of inquiry that the Foundation seemed to purposefully avoid.

The story of a rejected proposal would likely have been consigned to oblivion if it weren't for the role that Xolocotzi occupied during the early and defining moments in the history of the Rockefeller Foundation's work in Mexico. Xolocotzi had transcended the role of a mere trainee to occupy a higher administrative post and distinguish himself as one of the more knowledgeable members of the Rockefeller program. His extensive understanding of Mexico's botany and ecology, his observations of agriculture and its impact on the dietary, economic, and social aspects in peasant households, and a keen sense of how to meld these considerations into rigorous scientific projects eventually established him as one of the most prominent advocates for an alternative agriculture in Mexico.

By reviewing Xolocotzi's role in the conflicts surrounding Mexico's agriculture, one can discern the pattern and practice of the Foundation's work in Mexico. The Foundation's effort to increase yields was predicated on organizing a corps of young scientists trained to view agriculture in a very different way from their predecessors. Orchestrating this shared view of what was meant by "scientific agriculture" entailed a much broader mission than simply transferring technologies. It necessarily brought the Foundation into an even larger project: the production of knowledge.[3]

Beginning in the 1940s, the Foundation facilitated the training of what would number more than seven hundred Mexicans in programs reflecting the kinds of university curricula offered by land grant universities across the United States. While providing generalized courses in basic agronomy, these offerings focused on applying modern techniques to agriculture, including chemical treatments for plant diseases, designing fertilizer regimens, and integrating mechanical harvesting and processing.

The training of agricultural engineers in the coming years would prove to be at least as important as the spread of celebrated miracle grains around the world. These trainees provided the detailed knowledge for how to transform landscapes based on ancient forms of agriculture into the modern era. In a very fundamental sense, the trainees held the front-line responsibilty for applying the package of US agricultural techniques to Mexico: Prior to the Foundation's mission, Mexico lacked a group of professionals trained in how to apply these techniques so as to transform Mexico's agriculture into a modern production system based on the integrated application of fertilizers, pesticides, irrigation, machinery, and redesigned crops.

This corps of trainees would prove to be an essential force for transforming not simply farmers' fields, but also basic agriculture-related institutions.

From the 1950s onward, the young trainees supported by the Rockefeller Foundation began to assume key positions within important ministries, exerting influence on the formulation of policies to advance modern agricultural practices. These policies encompassed a wide range of initiatives, including the construction of dams, the expansion of loans and finance for a variety of inputs, the integration of plant production and processing with larger agribusinesses, and the coordination of support among an array of public and private institutions.

By the mid-1950s, the Foundation's técnicos or agricultural technicians had not only secured numerous government and university posts but also held many positions in various industries. Corporations that included Dow Chemical, American Cyanamid, Shell de Mexico, and Anderson Clayton, among others, were eager to take advantage of the education that young Mexican trainees had obtained under their US advisors. Support for commercialized agriculture often occurred with linking the work of trainees, experiment stations, patronatos, and agribusinesses. The Foundation's mission fostered many of such linkages, including assistance with identifying a specialist for Anderson Clayton, advice about where to locate a nitrogen plant in Sonora for the Commercial Solvents Corporation, nitrogen testing for Ciba Geigy, and arranging funding to support trainees at Pioneer Hi-Bred.

While the Foundation's scientists accepted the value of their work among international agribusinesses as a positive signal regarding the contours of their training program, it also caused them to worry about attaining a larger objective. Discussions within the Foundation agreed that unless their trainees were also channelled to positions in government, the program might never reach the point "where it will run itself" (Wellhausen 1956: 1–4). One of the major yet less articulated goals of the Foundation involved regenerating the personnel necessary to carry on the newly fostered relationships between patronatos, agribusinesses, selected fields in the agricultural sciences, as well as carefully defined techniques to support a specific type of free-market economy — that is, the capitalist economy practised by their northern neighbour.

The Foundation provided trainees with the means to transform the landscape from a largely new perspective, and this new perspective fostered a new understanding of agriculture. At a basic level, the transformation of how Mexico's agriculture was perceived marked a significant shift in the production of knowledge, especially within the country's agricultural research centres and universities. This new perspective replaced the post-revolutionary priority of serving the millions of peasants in the countryside with an emphasis on increasing production levels for major crops and agricultural commodities. Many in the agricultural sciences conflated these objectives, presuming that increased production linked with the expansion of markets would benefit humanity. An interdisciplinary focused inquiry might well have fostered a fuller examination of this assumption. Yet the Foundation's records suggest the promotion of something much closer to wilful ignorance than a quest for knowledge.

Even as the Foundation's work in Mexico was rapidly expanding, the program's leadership resisted confronting questions regarding the larger social and political consequences of its work. Carl Sauer's observations challenged one of the most basic principles in the Foundation's scientific mission: whether the Foundation actually possessed sufficient knowledge about the landscapes, people, and society to comprehend what was broken in Mexico's agriculture. Sauer's observations, however, were marginalized, perhaps owing in part to his being a non-scientist and therefore disregarded as an outsider to the Foundation's mission. Yet in the coming years there would be many others, including insiders, who would raise basic questions about the conduct of this scientific mission.

The fact that Xolocotzi had received Rockefeller Foundation grants supporting his advanced studies in the US and was employed by the program's Office of Special Studies meant that his critiques were those of an insider and therefore much more difficult to ignore. In 1956, Xolocotzi wrote a letter to Dr. Edwin Wellhausen, then director of the OSS, regarding an upcoming meeting that would include members of the Rockefeller Foundation and proposing discussion of a number of controversial issues:

> To emphasize the need to supplement the agricultural program with sociological studies which have as their main objective the clarification and presentation of the social tendencies and repercussions resultant of the technological advances achieved during the period in which the program has been in effect ... there is the possibility that this disequilibrium in the rapidity of development in these various factors might occur [sic] and lead to the nullification, for all practical purposes, of the gains obtained in the application of modern technology. (Caire-Perez 2016: 107–8)

As one of Xolocotzi's most knowledgeable biographers later noted, his memo came perilously close to identifying a more fundamental flaw in the Foundation's mission: "The partnership, in short, between the Mexican government and the Rockefeller Foundation had achieved its mission but neglected a concomitant factor of increased food supplies" (Caire-Perez 2016: 107–8).

The fact that Xolocotzi submitted such a letter to a man who served as a direct link between the OSS and the Rockefeller Foundation must have been shocking. Xolocotzi surely knew his counsel ran counter to many of the operating procedures accepted by so many US scientists in Mexico as well as many, if not all, of the Foundation's officers. Even more stunning is the fact that Xolocotzi advanced such a blunt assessment regarding a disequilibrium between rich and poor farmers and, even more damaging, questioned who the intended beneficiaries of the increased production were. His statement presaged so many critiques of the Green Revolution to appear in the coming decades.

Amid the hundreds of trainees supported by the Foundation over many years, it is important to compare their distinct trajectories, for they reveal the economic and political value of different scientific approaches. Xolocotzi, for example, moved quickly from his early work with the OSS to teaching what would be a substantial cohort of students in such fields as biology, botany, and agroecology at both the undergraduate and graduate levels. While possessing an impressive resume, with research extending from peasants' agricultural practices to the early work of the Rockefeller Foundation in Mexico to conducting seminars advancing the burgeoning field of agroecology, Xolocotzi nonetheless never gained the resources or political authority exercised by many of his brethren supported by the Foundation. Or was it his orientation as a researcher that would destine him, as one of his former students would observe, to a position of institutional powerlessness? (Trujillo Arriaga 1990).

By contrast, Efraím Xolocotzi's other colleagues, who later assumed significant influence at Chapingo, possessed far less intellectual prowess. Ingeniero Oscar Brauer Herrera, for example, who studied at Chapingo from 1945 to 1951, possessed a far more limited knowledge of the broader cultural, social, biological, and ecological contours of Mexican agriculture. Instead, his resume emphasized precisely those arenas for orchestrating what the Foundation had put in place: integrating the agricultural economies of Mexico and the US based on achieving high levels of production by utilizing capital-intensive infrastructure and commercial relations with larger firms operating in global markets.

It becomes clear that the Foundation's work in Mexican agriculture extended well beyond a simple mission of technical assistance. From the curriculum at Chapingo and other universities, to the movement of its graduates into positions of authority and influence in a range of government agencies, to the social relationships forged with commercial enterprises, all facets combined to fundamentally redesign the pattern and practice of

Mexican agriculture and transform its landscapes. In 1972, the director of an economic research institute in Mexico City described this process in stark terms:

> The multinational food processing firms ... act as monopolies, increasing the cost of food.... Determining the zones of production and the types of crops and deciding what is to be exported. They also determine what seeds, fertilizers, insecticides, and machinery should be used, and they fix the salaries of the field and factory workers.... In the broadest sense Mexican agriculture is victimized and controlled by foreign firms in the food industry.
> (Mejido 1972: 191)

The Issue of Alternatives in Science

In 1984 Xolocotzi reflected on the changes that had occurred in his nation's physical and intellectual landscapes during his lifetime. He noted that one of the legacies of the OSS was an emphasis on the technological aspects of agriculture, to the exclusion of socioeconomic considerations (Hernández Xolocotzi 1984).

Xolocotzi's reflections are important precisely because of his advocacy for broadening the scope of scientific investigations. His insistence on this from the very inception of the Foundation's work in Mexico was especially noteworthy; awaiting a time when "improvements" might reach the vast numbers of campesinos risked losing ecological knowledge that might never be recovered. Additionally, Xolocotzi appeared to be sensitive to what should have been obvious to everyone at the time: if the project of assisting Mexican agriculture did not *begin* with Mexico's campesinos, it might never include those tens of millions of households.

Xolocotzi's other significant role was to suggest an alternative scientific approach to improving agriculture. Proposing another path not only aligned with the advocacy of others inside the IIA, but it also reflected Carl Sauer's recommendation to avoid importing US-styled agriculture and John Dickey's cautionary memorandum that the negative consequences of the Foundation's work would soon emerge across Mexico. Combined, all of this provided more than ample opportunity to consider more carefully Xolocotzi's critique: Was there not another way to pursue a different research agenda where Mexico's campesinos would be beneficiaries instead of casualties of the process?

The exclusion of a campesino-centred program at Chapingo, however, was already integrated into the pedagogy advanced by the Foundation and its scientists. Once again, as one of those who interviewed and spoke to Xolocotzi at length described the situation among those working at the interface between Chapingo's researchers and campesinos: "While they [extension workers] sympathized with peasants' lot in life, extensionists neglected the possibility that campesinos might know what they were doing as farmers. Extension precluded any serious study of local knowledge and consideration for factors of the communities in which they worked" (Caire-Perez 2016: 42).

"They [extensionists] disclaimed any interest in the ecological, economic, social, political, and cultural matrix in which they diffused this technology … because of the development discourse they embraced" (Caire-Perez 2016: 60). The discourse of development at Chapingo, however, was dominated by the Foundation scientists.

In the coming years, Xolocotzi's own work would emphasize the importance of studying biology, botany, and sociology while noting the problems of the reductionist stance that dominated US agricultural science. In 1960 Xolocotzi spoke about the shortage of researchers in such fields as genetics, physiology, cytology, and ecology — but also the lack of the social scientists.

The science brought forth by the Rockefeller Foundation at Chapingo came to reflect a source of conflict within the sciences that would only grow in the coming years. One author, noting Xolocotzi's different orientation at Chapingo from others, would foreshadow what would demarcate a vast gulf between the Rockefeller's management of science and another alternative — agroecology — that would contain certain roots Xolocotzi's work: "While he understands the pride that colleagues, politicians, students, and foreigners may have taken in relation to the changes of Chapingo and its agricultural education, he saw fundamental flaws. The new laboratories, libraries, study halls, hybrid seed development, was leading the promotion of a model of agriculture incongruent with Mexico's reality" (Caire-Perez 2016).

Science and Agency

One need not ascribe a sinister or conspiratorial intent on the Rockefeller Foundation's actions in Mexico. Indeed, thinking about the Foundation as the centre of a conspiracy both misrepresents the nature of its activities in Mexico while failing to appreciate the extension of its work across the Americas. The difference might be characterized as comparing a small gang of criminals to a global network of organizations including tens of thousands of people. Where the criminal gang might endeavour to keep its work secret and remote from the public eye; an international network seeks public disclosures and support. A key distinction for the international organization is teaching everyone how to think about and approve of its work — the production of knowledge.

The record of the Foundation's team suggests the pursuit of a model reflecting a value for many across the Americas: greater production as a social goal, especially the production of food. Yet, the value placed on the production of goods — and especially their place in so-called science-based projects — was an arena of contested terrain. At issue were not simply matters of whose production and whose profit, but a broader field of adverse consequences, including hunger, unemployment, and ecological destruction.

One of the features of the Rockefeller scientific team was not simply the growth and valuation of all kinds of knowledge, but the dismissal of other knowledge. This feature, sometimes referred to as scientific reductionism, meant that even as data collection and analysis about productivity was taking place, knowledge about its broader consequences was continually being obliterated. At times, the resistance to examining the values in science-based projects resembled a wilful ignorance. All which raised a fundamental question: What were the larger consequences for the Foundation's management of science and its corollary projects involving the production of knowledge?

Even more powerfully, the Foundation put in motion a vision of agriculture supplanting the purposes of the Mexican revolution. The foundational purpose of a bloody struggle to throw off an oligarchy possessing virtually all wealth and political power had, by the end of the century, been displaced by a series of quasi-political coups. At the centre of one of these coups was a quiet mission of technical assistance that would not just remake agriculture and what counts as agricultural knowledge but once more marshall lands and power into the hands of the very few over the many.

The question "How has the increased production in Mexican landscapes fulfilled the promise of more food for all?" should indeed prompt a more fundamental discussion about the nature of the Foundation's scientific mission. If there was evidence pointing to the mission fostering inequality between rich and poor, including in access to basic foods, surely the Foundation must investigate this hypothesis. Xolocotzi's questions, echoing the observations of Sauer and Dickey, pointed to a consequence increasingly recognized by more and more people. For all of its achievements, the Foundation's work in Mexico was producing not just increased yields but increased social tensions.

Xolocotzi, of course, was suggesting certain course changes to both increase production and provide foods to those most in need. To accomplish these combined goals, he argued that the Foundation should place Mexico's peasants at the centre of its work. For Xolocotzi, Mexico's peasants could serve as both a vital reservoir of information about how to improve yields with both greater resources and the removal of obstacles. Xolocotzi's advocacy, however, presented a conundrum for the Foundation: how to remove obstacles when these included patronatos, agribusinesses, and others who were not intent on sharing resources with peasants seeking water, fertilizers, access to credit, and entry to so-called free markets?

Even though Xolocotzi numbered among the faculty at Mexico's major agricultural university at Chapingo, many more of its faculty members included US agricultural scientists and researchers working with the Rockefeller Foundation. "Hernandez later suggested that the strategically aimed money and collaborative efforts [an apparent reference to the Ford and Rockefeller Foundations] towards certain departments, gave impulse to a US-styled focus on learning, research and extension without an appreciation for the socio-economic context" of Mexico's farmers during the Colegio's inception (Caire-Perez 2016: 162).

The science brought forth by the Rockefeller Foundation at Chapingo came to reflect a conflict within the sciences that would only grow in the coming years. One author, noting Xolocotzi's distinct orientation at Chapingo compared to others, foreshadowed what would eventually mark a vast gulf between the Rockefeller's management of science and an alternative approach — agroecology — whose roots can be traced, in part, to Xolocotzi's work:

> While he understands the pride that colleagues, politicians, students, and foreigners may have taken in relation to the changes of Chapingo and its agricultural education, he saw fundamental flaws. The new laboratories, libraries, study halls, hybrid seed development was leading the promotion of a model of agriculture that was incongruent with Mexico's reality. (Caire-Perez 2016: 162)

Yet even Xolocotzi's words were too generous; it was a critique that failed to accurately identify who held responsibility or agency in the management of science and the production of knowledge. Once the MAP scientists had embarked on the path favouring those farmers with access to capital, irrigation, larger holdings, and linkages to larger firms and markets, a vital element in altering Mexico's landscape was set in motion. The most essential and corresponding step was to embed these values — particularly the linking of productivity with the private accumulation of capital — in the concept and practice of agricultural science. At this juncture, knowledge production became more fully integrated with the production of agricultural commodities ... while obscuring this purposeful connection and instead providing a seemingly neutral narrative for obliterating traditional agriculture and the campesinos whose practices suggested a potentially viable alternative to industrial agriculture.

In the coming years, however, extinguishing a centuries-old alternative to industrial agriculture would prove to be a much more difficult task than envisioned by the Foundation and its scientists.

Notes

1. La Escuela Nacional de Agricultura, established in the early 1920s, would in the early 1960s be transformed into a part of the system of national autonomous universities, becoming La universidad autónoma de Chapingo (UACH), both of which are referred to in this work as Chapingo.
2. For the purposes of this book, I refer to Dr. Efraím Hernández Xolocotzi simply as "Xolocotzi." This choice is partly for readability, but it also reflects how many of his students and colleagues at Chapingo addressed him — whether as Maestro Xolocotzi or simply "Xolo." The use of "Xolocotzi" also highlights, for many, a celebration of his Indigenous heritage. Readers searching for his publications or writings about him should also try variations of his name, including Hernández, Hernández Xolocotzi, and Hernández X.
3. A small sampling of authors who have influenced this text from a vast literature relating to the sociology of knowledge, the political economy of knowledge, and philosophy of knowledge in addition to others already mentioned in the first chapter (e.g., Fleck, Kuhn, Noble) includes Paulo Freire (1970), Michel Foucault (1982), Jurgen Habermas (1976), and Charles E. Rosenberg (1976).

5 Expanding a Green Revolution across the Americas

On January 1, 1961, Mexico's secretary of agriculture and the Rockefeller Foundation agreed to end the Office of Special Studies (OEE) with the formation of a newly reorganized Ministry of Agriculture — the National Institute for Agricultural Investigations (INIA). With the merging of the OEE and IIA, it might appear that the divided scientific missions and the conflicts between the campesino-centred agriculture versus the US industrial model were peacefully reconciled in the INIA. The resolution of these conflicts, however, was more stark: campesino-centred agriculture had been firmly defeated by an industrial model. As a representative of the Rockefeller Foundation noted, the leadership positions at the new government body were now dominated by Rockefeller scientists, most of whom had served in the former OEE (Wellhausen 1961).

Terminating the MAP offered an opportune moment for the Foundation to conclude its mission. In a tangible sense, the latter had achieved significant milestones: launching experiment stations, training a corps of professionals already dominating the ranks of many state bureaucracies, implementing industrial standards for cultivating, harvesting, and processing agricultural commodities, and forging strong connections between elite farmers and agribusinesses. Setting in motion a self-replicating model of agriculture that no longer required the Foundation's guidance and intervention constituted a major coup; in a little over a decade the Foundation had transformed Mexico's agriculture from the revolutionary paradigm promising to serve campesinos to a fully integrated project linking state resources and elite farms to global agribusinesses and markets.

Exporting the Model: From Mexico to the Americas

In 1951, the three scientists involved in founding the Mexico program co-authored a paper advocating for expanding the Foundation's work in Mexico to other parts of the globe, with particular emphasis on the Americas.

> There are many ways in which the Foundation could become more widely helpful in Latin America, but none that appears more immediately promising than that of exploiting on a wider front the success of its present agricultural activities. The Mexican Agricultural Program stands as a hub around which future developments can be built.... The Mexican program should serve as a training center ... for young Latin Americans destined to return to their own countries. (Bradfield et al. 1951: 11)

Establishing the network of international agricultural research centres (IARCs) is often presented as beginning with the founding of the CIMMYT in Mexico and the International Rice Research Institute (IRRI) in the Philippines. These two centres were soon joined by other IARCs established around the globe.[1] Early funding by the Ford and Rockefeller Foundations leveraged much larger contributions from the World Bank, United Nations agencies, sovereign wealth funds, national governments, and private sources. Following the Rockefeller MAP model, IARCs tended to focus on select food commodities and economic and trade policy.

While the IARCs are often portrayed as emerging in the 1960s, their foundations can be traced to discussions that began much earlier. In their 1951 memo to the Foundation, the original three scientists wrote about the potential for internationalizing the Mexican mission:

> It is conceivable that the type of operation which has been successfully carried out in Mexico and has extended to Colombia can be extended not only to other key Latin American countries but possibly that tremendous opportunities exist in such areas as India and the Philippine Islands and perhaps in Japan and at a later date in other Asiatic countries. (Bradfield et al. 1951: 11)

A more blunt characterization of the "problem," as they saw it, defined campesinos and their traditional methods as an obstacle to scientific agriculture. Such a narrative appeared in a a longer work the team published on the problem of supplying food on a global scale. In sending a team of two plant breeders (for maize and wheat) to evaluate problems in Colombia in 1950, the three scientists described the pivotal role of scientific agriculture:

> The Colombia of 1950 was a land of great riches and of abject poverty, of magnificence and of squalor, of culture and ignorance. Education was not compulsory and was not even available in many areas. Although about three-fourths of the urban population was literate, three-fourths of the rural people could neither

read nor write. Sunk in peasantry, they looked to political rather than scientific remedies to improve their lot. What Colombia really needed to improve agriculture and rural life in general was more science. (Bradfield et al. 1967: 218)

Embedded in the seemingly neutral prescription of "more science" was a formidable political agenda, rooted in the particular kind of training designed and offered by the Foundation's scientists at Chapingo. In a memo discussing the Foundation's success in Mexico and the possibility of extending its model to Colombia and beyond, the issue of how to manage the recruitment of even more scientists pointed to the special character of its science: "While the word science is spread around pretty thickly, as you see, what has to be done involves a lot of management in politics, and public relations — we must have in the organization the kind of people who know how to do it" (Weaver n.d.).

In 1951, the team of three worked with the Mexico program director, J. George Harrar, on writing a proposal to establish a new and separate division within the Foundation based on the MAP in order to "accomplish all of the objectives of a well-rounded program of international agriculture" (Harrar 1951: 8). The proposal made a series of recommendations, including one drafted by Bradfield that held significant importance for where and how scientific work in the Americas would take place.

> Since the improvement of agriculture is dependent upon improvements in education, health, transportation and the availability of machinery, fungicides, insecticides, herbicides, etc., priority should be given to the few situations where a well-rounded program of development seems most probably [sic] over situations which are not yet so ready for such an integrated program. (Bradfield 1951: 68)

The Bradfield memo, its support by his fellow scientists, and its approval at the Foundation's highest levels represents a rather unique archival record. In a simple sense it conveyed the elements of a plan for extending the Mexico program across the Americas. At a more profound level, it made explicit the political dimensions of its work. More than merely some neutral curriculum of technical assistance, the outline for conducting a program of "scientific agriculture" demanded constructing an array of social, political, and economic infrastructures. The requirements for ready access to machinery, agrichemicals, and transportation prioritized the favouring of major agribusinesses and unrestricted access to capital markets, international trade, and the political structures to ensure investment stability. It was, in short, a prescription to follow the Foundation's long-standing principle of "betting on the best."

In demanding such conditions, the Foundation and its scientists implicitly elevated the role of large landholders and agribusinesses while further marginalizing campesinos, the landless, and their households. Involvement in campesino protests over land titles, water rights, labour rights, and related political conflicts clearly fell outside of what the Foundation and its scientists regarded as their mandate, and such collaboration was a political alternative that they were unwilling to embrace. Decades later, the Foundation and international agencies would indeed turn their attention to campesinos and their agriculture, but the dye was already cast. The industrial model of agriculture now dominated the Americas. There was no place physically, politically, or economically for a competing model — certainly not one organized and directed by campesinos.

To understand more fully the operational side of the Bradfield memo, one need only turn to the generations of scientists that the Foundation trained throughout the Americas. These young men (and a very few women) would provide the wherewithal for extending the science of industrial agriculture.

Training a Generation of Técnicos

The history of expenditures supporting research, fellowships, and programs reveals that the Foundation had already begun to internationalize its work in the 1940s. From 1940 through 1960, the Foundation invested over $42 million in agricultural sciences globally, with nearly half allocated to Mexico and Latin America.

The major expenditures supported operating programs such as the MAP; fellowships and scholarships received approximately $7 million; and grants amounted to approximately $20 million. At the country level, the Foundation budgeted funding for national departments of agriculture, including ministries in Ecuador, Colombia, Chile, Guatemala, Costa Rica, and Honduras. For universities, Foundation support included programs and departments in Chile, Colombia, Ecuador, Guatemala, Brazil, Bolivia, and Peru.

By the early 1960s, many administrators of national departments of agriculture across Latin America had received advanced training at Chapingo or land-grant universities in the United States; the Foundation likewise supported advanced agricultural science training for many students within their departments. For the year ending June 30, 1963, the Foundation's fellowships and scholarships in the Americas included support for scientists in Argentina, Bolivia, Brazil, Chile, Colombia, Costa Rica, Ecuador, Guatemala, Honduras, Mexico, Nicaragua, Peru, and Venezuela.

By 1956 the Foundation boasted about the successful placement of 170 professionals it had trained in other parts of Latin America: "Already many of them have been advanced to positions of responsibility in ministries of agriculture, colleges of agriculture, research agencies and private industry; many of them are now helping in the training of a younger generation of technicians and scientists" (Rockefeller Foundation Annual Reports 1956: 180). During the period 1959–1973, the Foundation's funding throughout the Americas exceeded $4 million. Among the many scientists who previously had difficulty obtaining even a few hundred dollars, the Foundation's support, extending to thousands of dollars, represented a defining force regarding their research direction and agricultural demonstrations (CIMMYT 1966–1973).

The real numbers, however, do not adequately reflect the content and impact of this training across the Americas. The Rockefellers' scientific agriculture can be captured in both its reductionist character as well as its lack of consideration for other disciplines such as ecology, ethnobotany, or sociology. Indeed, the disruptiveness of Xolocotzi's presence and his being sidelined as a member of the faculty at Chapingo offered a stark lesson about where young trainees should not venture in their professional work.

Marginalizing what was perhaps most central issue raised by Xolocotzi — the need to evaluate the impacts of scientific agriculture on campesinos, particularly in light of the already evident growth of inequality — was a quintessential feature of the Rockefeller mission. The negative consequences of the Foundation's scientific agriculture were to be ignored, and if it begged questions about the apparent worsening of social conditions, this was the responsibility of others — non-scientists — to address.

It is once again worth considering the contrasting narratives about what was taking place in Mexico. The Foundation refers to tensions surrounding the gulf between science and technology on one hand and culture and society on the other. A more accurate accounting, however, points to the use of science and technology as devices for defeating laws and institutions (e.g., agrarian reform, communal and ejido property rights) intended to protect and advance the role of Mexico's campesinos. For many, the Foundation's success was nothing less than a counterrevolution against the aspirations of millions of Mexicans who had dedicated their lives to creating an alternative agriculture, a new set of political relationships favouring those on the bottom, and a new state.

Over fifty years, the Rockefeller Foundation, in conjunction with other centres of scientific investigation in the United States, transformed agricultural institutions throughout the Americas. Although there are more than 45,000 scientists with advanced training in the world, the 1,700 scientists employed with the Consultative Group on International Agricultural Research (CGIAR) have exercised significant power over agricultural research in many Asian, African, and American countries (Collinson and Wright Platais 1991: 2–5). As a group, the Green Revolution scientists command vastly greater salaries, access to governmental agencies, and maintain working relationships with an array of private-sector firms working in banking, finance, and trade. Their professional power, especially when linked with multinational agribusinesses, strongly influences the trajectory of national agricultural policies.

By the close of the twentieth century, the budgets for three of the international agricultural research centres operating in Latin America (the CIMMYT, CIAT, and CIP) exceeded the operating budgets for many of the largest agricultural colleges in the Americas. Between 1985 and 1989, the IARCs provided training to an estimated 25,000 scientists throughout the world.

This spreading scientific agriculture throughout the Americas would have an enduring legacy. First, it would further entrench commercial agriculture across diverse regions. Second, it would replicate a network of institutional relationships linking agribusiness, state agencies, and universities. Third, it would establish a model for generating expert, science-driven technological knowledge, shaping a conceptual framework for addressing agricultural challenges and their solutions. For decades to come, the Foundation's initial work would serve to marginalize millions of peasants throughout the Americas and beyond, creating sweeping barriers to sharing what would be the valuable contribution held by so many of these workers and their communities: their skills and knowledge for how to construct a livelihood that combined agriculture and ecology, sovereignty in food production, and integrated local and regional economies.

The training of técnicos in the principles of industrial agriculture amounted to a "colonization of knowledge," ensuring the perpetuation of a scientific-technical mission across the Americas while erasing consideration of campesinos, the adverse consequences of this model, or viable alternatives. As described years later by Peter Rosset:

> The loss of local knowledge is typically the result of decades of imposition of exogenous knowledge through projects, conditions imposed for credit and loans, technical assistance or agricultural extension, and ubiquitous advertising in favor of the pesticide-dependent monoculture model that characterizes the so-called Green Revolution. This supposedly "modern" knowledge has been pushed for years, to the detriment of the inter-generational transmission of peasant knowledge. Many members of the typical peasant population have not learned the traditional knowledges of peasant practices (which are mainly agroecological in nature), while others know something about them, but often devalue their own traditional knowledge and practices, considering them to be "backward" or "primitive," a source of shame. In other words, in this historical process of imposing a capitalist rationality, there has been a "colonization of knowledge," through the imposition of so-called "technical knowledge" that originates in the countries of the North and in agronomy faculties, and which, generally speaking, is not usually suited to the reality of peasant agriculture. (Rosset et al. 2025)

The Rockefeller Foundation would set in motion an array of negative consequences, including the violent repression of marginalized communities. It is vital to recall that the Foundation had been warned about the emergence of such negative consequences on many occasions over many years.

Even as IARCs were established across the world, questioning of their work was also moving apace. In a series of investigative reports issued by one of the research divisions of the United Nations, its director argued that the Green Revolution was making the position of large numbers of campesinos even more precarious, a situation arising by the very "terms of incorporation" into an exploitative market economy. The director concluded that the Green Revolution "replaced self-sustaining local production/consumption systems" with a reliance on a new economic relationship dominated by market-based institutions, including distant research centres.[2]

The very terminology the scientific and technological mission sought to marginalize millions of campesinos' traditional skills and knowledge as worthless. The imposed authority of scientists, tècnicos, and other experts as superior was precisely what Sauer, Xolocotzi, and others had cautioned against, recognizing that millions of Indigenous people and campesinos often possessed vast knowledge and wisdom about local ecologies and the provisioning of food, shelter, and medicine that formed the basis of their livelihoods, regional economies, and the protection and enhancement of natural resources.

By the twenty-first century, many experts began to realize the cost of losing this traditional knowledge: it had been essential to building resilient systems for ensuring local food production, restoring natural ecologies, and protecting forests, among other aspects. Yet by then, the missionaries of industrial agriculture had often succeeded in making campesinos and their communities doubt, and even feel ashamed of, their traditional knowledge (Rosset et al. 2025).

Notes

1. The reference to IARCs would eventually be folded into a more formal international network with the founding of the CGIAR. The CGIAR would serve as an institutional hub to coordinate the IARCs's work, particularly with respect to allocating financial support via international agencies as well as national donor programs.
2. Green Revolution was a term used by William S. Gaud, former administrator for the US Agency for International Development to refer to the package of technologies fostered by the IARCs and characterized by improved seeds, chemical and mechanical inputs, and infrastructures including irrigation, finance, and market regimes. See Lester Brown's *Seeds of Change* (1970) for a fuller explanation on the origins of this reference.

6 The Violence of the Green Revolution

John D. Rockefeller Sr. and his advisors saw agricultural production as critical to prosperity. At the beginning of the twentieth century, they embarked on an unprecedented effort to help poor farmers in the southern United States increase agricultural productivity. This effort profoundly influenced innovation in the agricultural sector in the United States and, later, in countries around the globe.

(Dr. Judith Rodin, president, Rockefeller Foundation 2013)

Despite the Foundation's claim of a long tradition of serving the needs of poor farmers, its foremost metric of success focused largely on a measurement more important to agribusinesses than rural households: productivity of a specific commodity on a unit of land. Considerations of a broader set of nutritional, public health, ecological, or other social consequences were largely, if not entirely, absent from its own progress reports. The Foundation's focus on production left a powerful and lasting impact on the scientific communities it supported, a legacy that I have elsewhere referred to as the production paradigm (Jennings 1984). Integral to shaping the research focus of scientists to expand their knowledge of the commercial aspects of agriculture, this paradigm encouraged the neglect of its larger social consequences.

As Professor Xolocotzi noted, the presence of social scientists was largely non-existent among the Foundation's senior representatives, who dominated the faculty and curricula at Chapingo, or among the generation of trainees who would populate agricultural agencies across the Americas. It should come as no surprise that by the early 1950s — as cautioned by Sauer, anticipated by Dickey, and observed by Xolocotzi — an array of adverse consequences was blossoming in Green Revolution landscapes. The linkage between this spreading conversion and social tensions was inescapable.

By the close of the 1970s, the authoritative series of studies sponsored by the United Nations documented the destructive consequences of integrating campesinos into larger market regimes.

> As the new seeds require chemical inputs, farmers must purchase them from industrial sectors ... these factors force the farmer to reorganize his economy as to be assured of producing a standard minimum surplus which can be marketed in order to pay for the purchased inputs and for any interest due on the loan needed to buy them. For most cultivators to be incorporated according to these terms, the consequences have been quite dramatic ... the new technology has forced millions to become credit-stricken entrepreneurs. Strength in the market and credit-worthiness become essential for living off and working the land. (Oasa 1981b: 44)

The reaction among many poorer agricultural communities was not passivity, however, but active rejection of the new terms of incorporation into market forces beyond their control.

Paralleling the production blast-off fostered in selected regions throughout Mexico, rural protests and other forms of resistance were appearing in many of these same areas. The Foundation's achievements in significantly boosting production were becoming increasingly linked with a concentration of wealth among an elite group of farmers. Contrary to the notion that increases in production would "trickle down" to benefit the larger society, the early consequences were pointing to another pattern: a rising violence accompanying the worsening lives of Mexico's campesinos (Pearse 1980: 114–15). Among the first signs of social unrest was the growing membership of workers in outlawed peasant unions.

The General Union of Mexican Workers and Peasants (UGOCM) formed in 1949, and while it lacked official government sponsorship, it attracted an expanding membership in the 1950s and 1960s as workers responded to deteriorating conditions in work and wages. It urged its members to engage in land invasions to call attention to the plight of rural people (Strategy for Small Farmers 1975), and in 1957 the UGOCM occupied the Cananea Cattle Company, one of the largest private holdings in Mexico. Again, in 1958, the union organized thousands of peasants in simultaneous seizures of land in the states of Sonora, Sinaloa, and Baja California. In each of these regions the most important targets were large commercial operations — ones owing the logic of their production to the Foundation's model of scientific agriculture.

Peasant activism spread during the early 1960s to other regions in Mexico. Ruben Jaramillo, for example, joined with five thousand peasants to invade pasture lands in Morelos (Strategy for Small Farmers 1975). Jaramillo, a commander who had served under Emiliano Zapata, continued his work in subsequent decades to secure campesino land access rights, to provide them with the means to cultivate these lands, and to for them to control the pricing of their harvests. Contrary to the narratives of rural peace following the revolution, Jaramillo and others working on behalf of Mexico's campesinos struggled against the increasing suppression of these demands. Jaramillo's followers, the Jaramillistas, were inspired by Emiliano Zapata and the campesino revolution. However, "[t]he government ultimately responded with repression, forcing the Jaramillistas into armed struggle and transforming their calls for local reform into a broader critique of capitalism" (Padilla 2008). In Chihuahua, peasants intensified their efforts by forming guerilla groups. Ramon Danzos Palomino led thousands of peasants in marches and land invasions across several states in central Mexico, and they received substantial support among workers in Puebla.

Between 1967 and 1968 at least a dozen land invasions occurred across the state of Puebla (Paré 1972). Discontent in campesino communities in Puebla and nearby states had been mounting since the early 1960s against political bosses or caciques working in conjunction with larger landholders, and by the close of the decade campesinos had coalesced into a region-wide movement. These regional protests culminated in August of 1973 with a generalized protest opposing increased taxation, especially on communally farmed properties (Ramos et al. 1984). For certain members of the CIMMYT's scientific staff, the social tensions arising among Puebla's campesinos offered a potentially useful opportunity for the Foundation to finally deploy its program of scientific and technical assistance to increase agricultural production among campesinos.

Within this context, the Rockefeller Foundation and its scientists at the CIMMYT embarked on a new project in Puebla, a Mexican state where poor campesinos working small plots of land dominated the landscape. Plan Puebla was specifically designed to demonstrate the advantages of modern agricultural technologies to subsistence farmers who relied on traditional maize varieties and older farming methods (CIMMYT 1974). After decades of "betting on the best," the launch of Plan Puebla in 1967 marked the CIMMYT's first significant effort to address the needs of Mexico's poorest agriculturalists, a response to mounting criticism that its mission had consistently overlooked the struggles of ordinary campesinos who lacked the capital, resources, and political connections to improve their place in Mexico's agricultural economy.

From its inception, however, Plan Puebla faced significant challenges, as the scientific team encountered an immediate credibility problem: none of its "improved" maize varieties performed as well as the traditional varieties cultivated by local farmers.[1] More critically, from the campesinos' perspective, these new varieties also failed in broader, practical terms, as many farmers found the improved maize to be unpalatable, prone to local pests, and unsuitable for producing the stalks needed for fodder.

The CIMMYT's emphasis on maximizing maize yields per hectare also overlooked a crucial question: what were the socioeconomic impacts of the program on the region's rural households? Plan Puebla's line of inquiry mirrored its earliest work with farmers across Mexico: evaluating technical and scientific assistance with respect to participating farmers and agricultural production on their fields. This production focus, however, ignored what was happening in other parts of the regional economy — especially the broader array of campesino households. Within a few short years, the CIMMYT had encountered a range of social-bound problems emerging among peasants in the region — particularly those campesinos excluded from participating in its "experiment."

By early 1973, the CIMMYT had terminated its sponsorship of Plan Puebla. Beneath the official narrative lay evidence of the Green Revolution's recurring "unanticipated consequences," particularly the negative social impacts that the Rockefeller Foundation had long sought to disassociate from its projects in "scientific agriculture."[2] The CIMMYT's termination of Plan Puebla likewise coincided with the arrival of a Ford Foundation–supported team of economists. One of their primary goals was to extricate the CIMMYT from projects like Plan Puebla. The broader lesson was clear: the CIMMYT needed to distance its scientific endeavours from the socioeconomic disruptions caused by the Rockefeller Foundation's earlier reshaping of Mexican agriculture while disguising its impact on social and economic relations across the Americas. Campesino-based movements throughout the state of Puebla continued to gather force in the years that followed. By early 1976, some ten thousand campesinos had gathered to protest government repression. In response, the government deployed an equal number of soldiers to the region and over a period of years it imprisoned leaders, broke up protests with military force, and reinstated coercive rule by caciques and larger landed interests (Garcia et al. 1984). For some observers, the fact that social conflicts continued to emerge across Puebla indicated that the model of industrial agriculture fostered by the Foundation was destabilizing Mexico's countryside in a way that no program of technical and scientific assistance could address. For others, it suggested something entirely different: that the legacy of the Foundation's work in Mexico had set in motion a model of agriculture that would ultimately displace campesinos from the land altogether.

An independent review of the Puebla Project by Luisa Paré analyzed the increased social conflict in Puebla accompanying changes throughout its agricultural landscape. Beyond evaluating crop yields and training programs, Paré and her colleagues documented how the modernization of agricultural production sparked new waves of conflict. These included tensions between private and communal landholders, owners of small plots and large estates, different types of wage labourers, and those politically connected to caciques, patronatos, partidos, or other elites versus those excluded from these networks.

But Paré and her colleagues were not alone. Those who documented the taking of tens of thousands of hectares of campesino lands in other states during the early 1970s noted that the growth of rural conflicts appeared to be a natural consequence of the further penetration of capital in the countryside (Bartra 1977). Other investigators analyzing a timeframe extending from 1977 to 1983 found that struggles over land in Mexico's northern commercial export zones surged from 217 to 534 clashes. Even carefully conducted studies analyzing the systemic state repression and violence that bled campesino organizations of their ability to oppose the growing inequality in access to land and resources could not fully convey the sheer brutality of the violence. Following nearly four decades of policies and practices to modernize its agriculture, the broader social consequences of the Foundation's project were being felt by campesinos across Mexico. One national peasant union decried the murder of one campesino every three days during 1984 and 1985 (Rubio 1987: 33, 192). In 1988, more than two dozen major episodes of agrarian struggle were reported by Indigenous communities in Chiapas, Oaxaca, Veracruz, Tlaxcala, Hidalgo, and Guerrero (Revista del Instituto Nacional Indigenista 1989).

Thus, the Foundation's scientific agriculture did more than forge new relationships between agribusinesses, patronatos, financial elites, and state institutions — it also fostered a climate hostile to campesino-based agriculture, its production practices, and its control over land and resources. The director of a series of UN-commissioned studies analyzing the impact of the Green Revolution summarized the vicious contradiction operating at the heart of Foundation-inspired projects across the world:

> Perhaps the most important aspect of the rural situation ... is the fact that the rich and poor do not simply co-exist. The accumulation of land by the rich creates a demand for labor which the poor are obliged to satisfy because of their land poverty or landlessness; moreover, the entrepreneurial success of the rich is made possible by the hunger and importunacy of the poor cultivator who is obliged to surrender his bargaining freedom and even pledge his future labor at a reduced price in order to sustain his family and meet current obligations. (Pearse 1980)

These dynamics underscored the observations by Dickey and others, who had counselled the Foundation to exit its entanglements and responsibilities for the obvious social conflicts that would result from its projects in Mexico's countryside. The explanation that the Foundation was simply engaged in the neutral science of agriculture would not hold for much longer.

The Rockefeller Foundation's departure from Mexico in the early 1960s coincided with less-publicized but significant adverse consequences that had emerged by the late 1950s. Between 1930 and the end of the 1940s the very small landholdings held by Mexico's peasantry had doubled to exceed more than one million parcels of land. As the Foundation's program took root in the 1950s, "they [the holdings of small, private farmers] were gradually reduced in number and size, with a substantial proportion absorbed into larger holdings after this period. By 1960 they operated on less than 4 percent of the land" (Chacón 2021: 44).

An equally chilling statistic captured what these changes meant for Mexico's peasantry: "By 1950, there were 1.5 million landless workers, who made up 30 percent of the rural population. By 1960, this number increased to 3 million and 55 percent of the population" (Chacón 2021: 44). Adding to this portrait of many rural regions in economic free-fall was the fact that between 1940 and 1970, Mexico's economically active rural population fell from 65 percent to 39 percent (Chacón 2021: 45). Missing from these images were the largely invisible workers, numbering in the hundreds of thousands, who annually made their trek to cross the US border to toil temporarily in agricultural areas of California, Texas, Arizona, and farther afield.

Beyond Mexico

The suppression of peasant organizations, labour organizers, and rural resistance was a growing feature accompanying the arrival of "scientific agriculture" in many regions of Mexico, Central America, Colombia, and beyond. Swiftly mounting evidence gathered by researchers around the world challenged the legacy of programs modelled after the Foundation's work in Mexico. Prominent in the themes among a rapidly expanding universe of critical studies was the evidence that one of the principal byproducts of the Green Revolution globally was to generate violence in the countryside by further marginalizing poor farmworkers and their communities.[3]

As Luisa Paré, Roger Bartra, and their colleagues were analyzing the social tensions arising from Mexico's situation, their findings on the negative social, economic, and environmental consequences were echoed in penetrating analyses by Vandana Shiva, Andrew Pearse, Cynthia Hewitt de Alcántara, and many others from Asia, Africa, and Latin America. To better understand the commonalities of the MAP and its consequences for other parts of the Americas, the work of Michael Taussig and his studies of Colombian agriculture are especially striking.

By the 1960s, a new generation of trainees supported by the Rockefeller Foundation were replicating Mexico's agricultural program in Colombia. Like their Mexican colleagues, the Colombian técnicos were making headway by integrating agricultural techniques — improved seeds, irrigation, mechanization — with agribusiness, finance capital, and international markets. As with Mexico, the projects fostered by the Foundation's work were transforming production in many landscapes. The issue as to how such changes related to the lives of those working in the fields, however, remained too often largely unexamined by the new generation of técnicos trained at Chapingo.

In regions like Colombia's Cauca Valley and the Atlantic coast, tensions tied to export agriculture had been brewing for decades. By the 1930s, violent strikes and agricultural plagues had forced companies like United Fruit to suspend operations. However, Colombia's investment climate became attractive again by the 1950s, aided in part by the Foundation's initiatives. Taussig describes how modern agriculture redefined traditional peasant farming relationships:

> Far from being a "backward" form of labor organization, the contractor system seems to intensify in concert with the development of large-scale capitalist farming. Like the sugar plantations of the Cauca Valley, for example, the banana plantations of the United Fruit Company on the Atlantic coast began by owning all land worked, with a directly employed work force. Driven out by violent strikes (and plagues) in the 1930s, the company returned in 1959, but now all work is done by contracted labor, much of which retains links with peasant farming, and the land is in the

possession of large Colombian landowners financed indirectly by the company through Boston banks. This new system protects the company from labor conflict, nationalization, and price-falls. Similarly, the rapidly growing cotton plantations depend totally on contracted seasonal labor for the harvest, which as of 1970 amounted to over 200,000 workers. Most of the other new commercial crops also depend on contracted labor (just as the US oil companies did). (Taussig 1978)

This reorganization of labour and land ownership, backed by powerful interests and laws, shielded agribusinesses from traditional vulnerabilities of labour unrest, state intervention, and market instability while deepening the marginalization of campesino communities across the countryside.

Creative Destruction

As the global critique of the Green Revolution intensifed through the 1970s and 1980s, one focal point of scrutiny was the wretched working conditions endured by vast numbers of migrants working in California's agricultural fields, many of whom were Mexican peasants displaced by the Green Revolution and forced to seek work as field labourers.

As Californian farm labourers sought the same rights as organized labour in other economic sectors, they were joined in their political struggle by many urban allies. The broadening of their struggles also led to a growing recognition that agricultural chemicals were not simply injuring and killing fieldworkers, but also affecting the larger public with exposures to harmful residues on fruits and vegetables, polluted drinking water, poisoned soils, and toxic air contaminants.

Another manifestation of the Green Revolution, which expanded by the 2000s alongside the growing drug wars, revealed another dimension of the Foundation's push to spread market forces across the Americas. With vast regions of the Americas already integrated into market economies, marginalized campesinos found themselves increasingly drawn into drug cultivation. Fields once dedicated to meeting local and regional food demands were now repurposed to serve the global demand for narcotics. While earlier local economies were rooted in the constraints of their ecological resources, water availability, and the needs of poorer communities, the new economy opened these regions to external demands that extended far beyond their borders.

The Foundation's Green Revolution now introduced fresh conflicts into societies caught between the forces of capitalist "creative destruction" and emerging efforts to address the ecological crises exacerbated by these same market dynamics. As the narrative of a Green Revolution faced increasing challenges, the contradictions of the Rockefeller Foundation's project in Mexico and the Americas began to emerge, including a central question: What would happen with campesinos, rural workers, and their communities?

For the Foundation's economists, the displacement of peasants was simply a part of creative destruction; the process of modernization in which older, "outdated" skills gave way to modern practices that aligned with global economies based on free trade and competition. Campesinos and farmers who were early adopters of these modern practices would advance, while others would learn the necessity of changing their skills to match the requirements of a new economy. The Foundation and many of its supporters contended that it was the responsibility of national policies to assist in the transition of peasants, rural workers, and their communities through programs extending from nutritional support and health care to education and training.

Dividing responsibilities between the Foundation and governments — one based on disseminating a neutral science, the other based on extending social welfare — facilitated a broader narrative in which those in the private economy — agribusinesses, finance capital and corporations — operated freely without any responsibilities for their consequences. When rural communities began to protest and fight against what was happening, it was the role of government to control the rising tide of social conflict. Indeed, this largely describes the period of the latter part of the twentieth century, when rural violence became more commonplace as the Foundation-modelled initiatives moved beyond isolated pockets of modernized farms to areas more densely populated by small farms, landless peasants, and the rural poor.

The "disequilibria" noted by Xolocotzi and many others became more pronounced with the spread of newly engineered seeds devised by the Rockefeller scientists. The seeds of change introduced by the Foundation throughout the Americas extended the logic of market economies. In Mexico this meant displacing campesinos and the promise of Zapata to control the lands they worked with a new set of principles based on the logic of faraway markets.

Xolocotzi was among the first to challenge the notion that the peasantry lacked valuable skills. More than simply a romanticized notion associated with rural workers, the deep knowledge possessed by campesinos was built on hundreds to thousands of years of designing an agriculture in the context of the surrounding ecology. The Green Revolution had introduced fresh conflicts into societies caught between the tension of global demand to creatively destroy their traditional economies according to capitalist logic and the new realization of the value of essential biological and scientific efforts to remedy a global ecological crisis fostered by these same market forces.

By the 2000s, the legacy of the Rockefeller Foundation's transformation of Mexican agriculture extended well beyond a mythic claim of simply producing more food for a growing population and helping poor farmers. The Foundation's historical reliance on market forces, from the efforts of Seaman Knapp in the southern United States to Norman Borlaug and the empowerment of Mexican patronatos in the Pacific Northwest contained a potent political agenda: authority over decisions about what to plant, where to harvest, and who would profit from an agriculture would be vested in free-market forces.

A growing chorus of critics understood that the Foundation's technical program based on introducing modern seeds did not exist in isolation but carried with it an entirely new social, political and economic context. In the 1940s and 1950s this meant forging a new set of social relations based on solidifying relationships between US finance capital, agribusinesses, and markets with Mexico's elite farmers, growers' organizations, and luxury markets.

In conjunction with these activities, the Foundation played a major role by transforming whole parts of the Mexican government to respond to the demands of dominant interests and market forces on both sides of the border. By the early 1990s, one manifestation of this convergence of economic interests emerged with a change to Article 27 of Mexico's Constitution: a law prohibiting the sale of ejidal and communal lands. This provision, which had previously protected public and collective ownership of land for millions of campesinos, was effectively voided in favour of private — including foreign — ownership. As one observer noted, the changes to Article 27, when combined with a linked international trade treaty (to be discussed in Chapter 12), resulted in the emptying of thousands of small villages of workers, most of whom would trek with millions of their countrymen to work as agricultural fieldhands in the United States (Wright 1991).

Additionally, in actively shaping the production of knowledge and training, the Foundation demanded that the state, including its public expenditures on physical infrastructure, laws, and regulation, respond to the demands of dominant market forces. These achievements, however, subjected many millions of Mexicans to the incredibly harmful social, environmental, and public health consequences of living under the rule of free markets.

Notes

1. Helen Anne Curry (2022: 204). In this passage Dr. Curry cites CIMMYT's head of its Economic Unit, Don Winklemann, as well as CIMMYT's Plan Puebla as primary sources.
2. Bruce H. Jennings (1988). See Chapter 7 for a more detailed analysis of CIMMYT's Plan Puebla.
3. Vandana Shiva's (2016b) work on this topic, *The Violence of the Green Revolution*, serves as the title for this chapter precisely because it so well documents and captures what took place in India, another major setting for the Green Revolution, and advances the argument presented in this chapter.

Part II

Critiquing a Dominant Science and Its Larger Consequences

Even as the Rockefeller Foundation's mission spread across the Americas, by the early 1960s, a growing critique of industrial agriculture had emerged across the United States. It was especially in California that these critiques fuelled an incipient movement focused on protecting the environment, with particular concern for the consequences of industrial agriculture, especially the hazards posed by agrichemicals as contaminants in food, air, and water. As this movement gained support to protect agricultural workers, California became the epicenter for legal challenges to the power exercised by agribusiness and petrochemical corporations. Amid these conflicts, a group of scientists at the University of California, Berkeley, earned notoriety for not simply their critique of industrial agriculture, but their advocacy for an alternative science, agroecology.

Among the questions that emerged during this period, one stood out foremost in the minds of many consumers: "What is the price for protecting the environment?" This was followed by other concerns, including "What are alternatives to the existing food system?" and "Does it make more sense to fix the problems surrounding industrial agriculture, or is it more compelling to pursue alternatives?"

Ecology, Chemistry, and Conflicts

By the closing decades of the twentieth century, California's agricultural landscapes contained two defining features. First, the state's agriculture was dominated by highly capitalized agribusinesses. Second, this model of industrialized agriculture confronted a growing tide of public condemnation, particularly for its damaging effects on workers and the environment.

The University of California's nine campuses provided a juncture where two divided perspectives on the state's agricultural approach collided. One group of researchers argued that the state's chemical-intensive model of industrial agriculture should be subjected to much greater regulations. In contrast, a larger group of industrial agricultural scientists argued that fine-tuning the application of agrichemicals would suffice.

It was in this context that Miguel A. Altieri came to occupy a prominent position with the University of California at Berkeley. Professor Altieri's role in this conflict reflected an intellectual journey that captured perfectly the linkages between California and Chile, along with his paradoxical transformation into one of industrial agriculture's harshest critics:

> It was not until I arrived in California as a young Assistant Professor at the University of California, Berkeley, at the end of 1980, that I realized that all my training in agronomy at the University of Chile was based on the California model. Most of my professors had done their PhDs at UC Davis, sponsored by scholarships from the Ford Foundation, and had returned to Chile to disseminate and perfect the horticultural-fruit production systems they had been trained in. Nowhere in my undergraduate education was there mention of the social and ecological impacts of such modernization schemes. The campesinos of Chile, their production systems, and their importance for national food security were totally absent from the curriculum. (Altieri 2020: 1)

Dr. Altieri's reflection was particularly telling as a critique of industrial agriculture, paricularly, as he explained, because his education in Colombia opened the door to discovering a radically different agricultural model based on campesino knowledge and ecological principles:

> The 1973 military coup changed my life forever as I had to flee the Chile of Allende and Neruda. I never felt so inadequate professionally than when I arrived to do my Masters at the Universidad Nacional of Colombia in 1974, where for the first time I was exposed to the intercropping systems used by small farmers on the slopes of that biodiverse country. My specialized monocultural training did not serve me well to understand the complex interactions underlying such systems. In addition, I saw farmers "sponsoring" weeds in their fields as opposed to completely eliminating them as I was taught. (Altieri 2020: 1)

He was soon working with others who encouraged him to establish research plots where beans could be grown amid weed complexes to assess the response of the main insect pest to weed diversification and other such relationships. This led him to an uncommon source of knowledge in the natural sciences: "I became interested in better understanding the experiential agroecological knowledge of farmers as a necessary component to develop[ing] a more sustainable agriculture" (Altieri 2020: 1).

Among the notable elements of Altieri's awakening to agroecology was his work with a researcher at the International Center for Tropical Agriculture in Colombia (also known as CIAT, one of the sister institutions of the CIMMYT). "While at CIAT I met entomology professor Ivan Zuluaga from the Universidad Nacional de Colombia in Palmira, who was actively involved in anti-pesticide campaigns and the promotion of biocontrol alternatives. In 1976 I joined Professor Zuluaga to co-teach a course that they named 'agricultural ecology.'" As Altieri later explained, it was in that course that he made his first attempts to advance what would become agroecology (Altieri 2020: 1–2).

While this dialectic of industrial agriculture reflected the earlier observations of Sauer, Xolocotzi, and others, Altieri was about to join a much larger team of faculty, researchers, and activists determined to challenge industrial agriculture. It was no small irony that precisely the kind of agricultural knowledge that was obliterated by the industrial model was slowly being recognized as an invaluable source for building its alternative. That one of the principal places for encouraging an alternative occurred in California might be explained by the fact that California was quickly learning about the contradictions embedded in its own model. Among the earliest and most alarming developments of the 1960s was a growing public concern surrounding one of its central features: massive applications of synthetic pesticides and evidence of their cascading harms.

Industrial Agriculture: Conflicts over Chemical Consequences

While global awareness of the negative consequences surrounding Green Revolution projects gradually expanded, skepticism regarding the science supporting industrial agriculture grew. Such criticisms began to challenge what the Rockefeller Foundation, US land grant colleges, and IARCs meant by technological and scientific assistance. More broadly, these challenges revealed differences within scientific communities across the United States, sparking debate regarding the nature of their research and the purpose of their scientific missions.

One of the places where an early critique of prevailing scientific projects at US land grant colleges held sway was at the Berkeley campus of the University of California. By the early 1960s, a gathering of professors within the biological sciences — a group known as the Division of Biological Control — had already identified itself as hostile to one of the pillars of agricultural research and the Green Revolution: the application of synthetic pesticides and related agricultural chemicals.

More than simply individuals opposed to agricultural chemicals, Biological Control researchers had accumulated a long and impressive record of accomplishments in California over many decades. Beginning with the successful control of a highly destructive citrus pest in the latter part of the nineteenth century, they soon made substantial contributions to the non-chemical control of pests in grapes, tomatoes, olives, ornamentals, and weeds. The early research in biological control was based almost entirely on the classical approach of foreign exploration, release, and colonization of parasites and predators for the suppression of agricultural pests. A fundamental disciplinary foundation of research supporting this approach was applied ecology.

The popular portrayal of biological control is as a method of pest control that relies on "good bugs to control bad bugs." For professionals in the field, the endeavour involves understanding complex predator-prey, parasite-host, and herbivore-plant relationships. As a technique for controlling insects, it is a non-chemical method that depends on extensive scientific investigation and ecological field studies that include behaviour, habitat characteristics, population dynamics and modelling, nutrition, and physiology. Despite a wealth of successful applications in fields across California and the nation, biological control was largely marginalized by green revolutionaries and land grant colleges with their focus on chemical controls.

The idea of a different kind of scientific revolution was sparked by young biologist Dr. Rachel Carson in the early 1960s. Dr. Carson, who had previously worked largely in obscurity, became one of the most influential scientists of the twentieth century with the publication of her book, *Silent Spring* (1978). A primary theme of her work focused on the broadly destructive consequences of synthetic pesticides, an essential part of US agriculture and a central feature in the model of industrial agriculture popular among many of the Green Revolution scientists. While many agricultural scientists dismissed Dr. Carson's analysis, *Silent Spring* was instantly celebrated at many universities across the US, especially in California.

The success of classical biological controls captured the very essence of how one might begin to construct an alternative to the Rockefeller Foundation's Green Revolution. For Dr. Carson, biological control represented a prime example of how to construct a science of agriculture based on thriving ecologies instead of poisoned landscapes.

> Examples of successful biological control of serious pests by importing their natural enemies are to be found in some 40 countries distributed over much of the world. The advantages of such control over chemicals are obvious: it is relatively inexpensive, it is permanent, it leaves no poisonous residues. Yet biological control has suffered from a lack of support. California is virtually alone among the states in having a formal program in biological control, and many have not even one entomologist who devotes full time to it. (Carson 1978: 292)

Within the praise that Carson reserved for California, one group was of particular importance: the Division of Biological Control at Berkeley, whose members were, in many respects, among the most vigorous proponents of an ecologically based agriculture. The Division became lauded, however, not merely as critics, but as a group whose successes were predicated on devising an agriculture premised on a healthy ecology — which operated at cross-purposes to industrial agriculture's widespread application of synthetic pesticides.

For the Berkeley faculty working with the Division, conventional farming practices represented a barrier to the successful application of biological controls. They noted that the problem was not isolated to pesticide hazards but extended to the larger issue of conventional agricultural practices. Conventional reliance on cultivating a single crop (monoculture) produced inhospitable settings for ecologically based farming methods, such as the use of natural predators to control damaging insect populations.

Impediments to the widespread use of biological controls were not simply located in field practices but extended to inhospitable research settings. Despite the Berkeley Division's successful application of such methods, institutional support for this approach dwindled as monocultures and pesticides increasingly dominated California's landscapes.

By the late 1960s the Division's relationship to California's agribusinesses had grown quite antagonistic. In 1970 four of the Division's faculty members — Drs. Donald Dahlsten, Richard Garcia, John Laing, and Robert van den Bosch — published a workbook with a foreword penned by Barry Commoner, a well-known biologist and activist in the United States.[1] Their work represented an unrestrained critique of pesticides, offering a final word urging political action: "it's your world ... don't leave it to the experts" (Dahlsten et at. 1970), a thinly veiled reference to the ongoing and profound conflicts of interest among a vast number of professors in the agricultural sciences and their ties to industrial agriculture. More than simply differences of opinion, these statements reflected a fundamental divide regarding the production of knowledge and evaluating its applications in the world. These debates likewise underscored the revolutionary — as well as counterrevolutionary — nature of science and technology in the world.

If pesticide manufacturers were unhappy with many of the earlier researchers who comprised the Division of Biological Control, Robert van den Bosch raised even more attention with his book *The Pesticide Conspiracy*, which launched an even more explicit attack on the influence of the chemical industry in academia. A sampling of his work illustrates why both industry and its associated faculty members were furious with van den Bosch:

> By and large, the aggie colleges (US land grant colleges) and their associated experiment stations and extension services are social anachronisms that view their mission as one narrowly oriented to crop production and agri-business and hardly concerned with broader social interests. What else explains their virtual neglect of the concerns of the farm worker, the consumer, the urban homemaker, and the environmentalist? This narrowness is perhaps explainable in largely agricultural states, where the universities are dominated by farming interests. (van den Bosch 1978)

The Division's notoriety did not end with van den Bosch's indictment of corporate farming or the university's failure to pursue more socially beneficial alternatives. Various of his colleagues were inspired by his words and shared his sentiments that novel forms of research needed to be pursued at the Berkeley campus. Over the next decade, many of the Division's faculty took part in activities including actively supporting farmworker causes, coordinating classes and seminars with public interest groups, and especially promoting alternatives to synthetic pesticides throughout the state. As a research group supporting public initiatives, however, they were viewed with disdain by many others within the university who possessed greater political power and access to substantial resources.

Even as the social and political significance of biological control remained largely invisible to the larger public, its value was occasionally noted by policymakers in the United States. In the late 1980s, for example, the congressional Office of Technology Assessment (OTA) reported that federal programs for the protection of biological diversity were essentially non-existent. Writing for a national legislative audience, the OTA examined one of the central contradictions between federal policies supporting the chemical practices of industrial agriculture and the need to maintain a healthy ecology: "They are aimed at recovery of species that are already on the verge of extinction rather than at prevention of future losses" (US OTA 1987: 64). The congressional study noted that regions with relatively small agricultural holdings and a variety of crops frequently provided a landscape able to support natural enemies of crop pests and a greater likelihood of supporting species and varieties that resisted disease outbreaks. While considerable support had gone to large-scale farming units using modern machinery and agricultural chemicals for the production of a single commodity, "As yet, relatively little scientific effort is being made to determine how biologically diverse farming could be made more profitable" (US OTA 1987: 64). The implication was clear: the continuing loss of agroecosystem diversity in the United States and throughout the world reflected its seemingly negligible value for economic development and correspondingly low value as a research priority. The rift between the scientific communities immersed in the use of synthetic controls and those whose work depended on biologically diverse farming systems reflected fundamentally different conceptions about their missions.

The challenges faced by the Biological Control faculty and their students at UC Berkeley paralleled, in many respects, those encountered by Xolocotzi and his students at Chapingo. Both groups sought to emphasize the vital importance of ecology as fundamental to advancing a healthy and resilient agriculture. At the same time, they experienced the hostilities of a much larger group of faculty whose research, expertise, and resources were often closely linked to the financial and political power of an array of commercial interests. Among these were agrichemical manufacturers and petroleum companies, both deeply invested in perpetuating the US model of industrial agriculture in Mexico, the United States, and globally.

The struggle between an agriculture embedded in the petroleum industry and its ecologically based alternatives would become a pivotal conflict. Before turning to the broader challenges faced by agroecologists, though, it is necessary to understand another facet in the landscape of chemicals and conflicts: the role of those working the fields.

Note

1   Barry Commoner, a biologist and professor, would be recognized as a leading voice in the creation of the US environmental movement and known for his work on ecology. Among his many publications, *The Closing Circle* was one of his best known works — and it was published around the time that he wrote the foreword for the book by the Biological Control group at Berkeley.

8 Fights in the Fields

About the time that Rachel Carson was revealing the ecological hazards of pesticides, another group of scientists had begun parallel work focusing on their human health impacts—specifically on the dangers faced by farmworkers. Indeed, only a few hundred yards from the offices of the Division of Biological Control at UC Berkeley, a group of public health scientists were assisting with the development of what would become some of the world's most restrictive pesticide laws to protect farmworkers and others.[1] The questions shaping much of the early work for this group of health scientists originated with what was becoming a focal point for so many Californians: farmworkers' demands to organize a union to represent their legal rights in the workplace.

For the many hundreds of thousands of Mexico's campesinos who had been forced from their communities to seek livelihoods elsewhere in the world, California's growers had communicated their eagerness to employ this army of dispossessed labourers. In contrast to the national labour laws forged during the 1930s establishing workers' legal rights and protections, including the right to organize and form unions, employers of migrant agricultural workers were largely exempt from such laws and generally treated these workers as people without legal rights. Mexican, Filipino and other migrant labourers across the United States endured a variety of abuses — inadequate housing, low wages and wage theft, little or no access to health care, insufficient and expensive foods, rarely available legal assistance, non-existent child care, and dangerous working conditions, especially due to pesticide exposure.

In the mid-1960s, Californians were awakening to a new reality: a cornerstone of their food production system — pesticides — posed a genuine hazard to farmworkers. The image of one of the nation's rising political figures, Robert F. Kennedy, sitting with César Chávez, the well-known farm labour organizer, instantly became a rallying point for broad public support for the creation of the United Farm Workers union. Growers who for years had simply rejected any form of negotiating with labour began to encounter an expanding number of consumers boycotting their products.

The fights in the fields, however, were only beginning. As growers were grudgingly forced by public pressure to negotiate with farmworkers and their union, the fights in the fields had set in motion greater scrutiny regarding the negative consequences flowing from industrial agriculture. While such scrutiny spread across university campuses, it was unevenly received by different disciplines. Even though the orthodoxy of increased production prevailed in the agricultural sciences, faculty in other disciplines and programs began to examine the broader array of negative consequences of industrial agriculture. For groups of sociologists, their investigations took the form of detailed accounts revealing the worsening social conditions for communities living in the midst of it. Certain natural scientists investigated the collapse of habitats and decline in species. Law schools sponsored legal clinics to assist farmworkers claiming law violations and damages. Physicians began reporting more broadly on worsening health conditions affecting farmworker communities. Occupational hygienists began to question older assumptions about "safe levels" of pesticide exposure. All of this began to accumulate and promote a larger public discussion about California's celebrated model of agriculture.[2]

By the early 1980s, awareness of farmworkers' struggles had likewise spread to larger coalitions of organizations exploring legal avenues for altering methods of industrial agriculture. The older notion that decisions about agriculture should be left solely in the hands of growers, agribusinesses, or free markets was being challenged across the state. For many advocates of farmworkers' rights, the fights in the fields began by documenting the hazards posed to farmworkers from pesticides. Because it was a legal tactic dependent on scientific evidence, central features of the political debate moved to the scientific realm — where different scientific perspectives assumed greater importance for commercial and financial transactions, farmworkers and their communities, and policies supporting industrial or alternative types of agriculture.

Industrial Agriculture and California's Poisoned Workers

With the enactment of a series of laws in California restricting the sale and use of pesticides beginning in the 1980s, chemical manufacturers pursued a legal strategy of placing the burden of proof on governments to demonstrate conclusively that specific pesticide ingredients could be proven to cause specific harms to human health. For corporate lobbyists, many toxicologists served a vital political role: to continuously pursue arguments about scientific proof of harm for specific exposures. In contrast, public health scientists whose training was based on a principle of "doing no harm," advocated for precautionary laws designed to avoid pesticides where the weight of evidence indicated a likelihood of harm. In judicial, legislative, and other fora, it became commonplace to witness this divergence in scientific approaches being recruited to achieve distinct political purposes — a divergence reflecting the earlier conflicts between scientific approaches in Mexico's transition to an industrialized model of agriculture.

The criticisms of industrial agriculture extended to many other campuses of the University of California. With the fuller emergence of farm labour protests and union organizing in the 1960s, faculty working at university programs in public health, law, social work, labour, rural studies, economics, and other disciplines began to coalesce into critical voices concerning a range of conventional agricultural practices and the treatment of agricultural labourers. Many of these same voices in academia, coupled with advocates on behalf of farm labourers, sponsored a variety of proposed laws — legislation that would gradually regulate an increasingly wide range of production practices on California's farms.

The imposition of ever-harsher regulations on agricultural practices largely lacked a clear vision for viable alternatives, as the public did not possess the level of legal awareness or legitimacy to create another kind of agriculture. Partly owing to the "freedoms" enshrined in the US Constitution, including the sanctity of private property, the only seemingly viable path throughout the close of the twentieth century and extending to the present has been campaigns based on consumer preferences, not rights as citizens. The result was that consumer demands for pesticide-free, organic, or other types of products were largely appeals to dominant market forces to offer specific kinds of products, frequently serving only niche markets and wealthy consumers. Few, if any, of these efforts have served as a basis for transforming the production practices and techniques of industrial agriculture.

By the close of the 1980s, legal efforts to aggressively limit the sale and use of pesticides had forced a more basic question into the open: What might an agriculture without synthetic pesticides look like? Instead of consumer demands, these efforts focused on workers' and citizens' rights to control hazards posed by pesticides.

This legal approach, advanced by labour unions at the beginning of the twentieth century, encountered deep resistance among larger agribusinesses across the United States, and especially larger growers in places like California's central valley. Beginning in the 1960s and extending well into the twenty-first century, unions allied with farmworkers experienced continuous conflict with powerful agribusinesses in California. Throughout the

1980s, farmworkers in California would prevail over agribusinesses, achieving important workplace protections. Despite these victories, however, the general public and even legal activists had little in the way of experience to imagine a different kind of agriculture. At the same time, private interests supporting industrial agriculture worked furiously to defeat any and all attempts by California voters to redesign the sector.

A pivotal political moment occurred toward the close of the 1980s, when California's voters were given a chance to enact a series of laws discouraging the use of fossil fuels, pesticides, and agricultural chemicals. The vast expenditures by chemical and petroleum companies, including several spawned by the founder of the Rockefeller Foundation, overwhelmed and defeated the efforts of farmworkers, environmentalists, and other groups.

Fields of Knowledge and Conflicting Claims

Beginning in the 1980s, California's legislature was crafting a series of laws undermining the principle that agricultural industries should be allowed to operate however they chose. The fights that began in the 1960s in California's fields over the denial of workplace rights for fieldworkers had, by the 1970s and 1980s, spread to a broader range of issues, including impacts on their communities. Public concern about farmworkers' exposure to agrichemicals awakened larger concerns among urban dwellers about contaminants in air and water as well as residues in food.

Adopting a response used by related industries, the agricultural sciences frequently responded by pursuing campaigns of reductionism, often characterized by making technical tweaks at a finer level of detail while ignoring larger patterns. In this way, industry lobbyists advocated for exhaustive research about the chemical composition of particular pesticides, the specific routes of public or worker exposures, the determination of harmful exposure levels, and other information gathered around finer and finer detail. Along this path, the industry would be able to pursue seemingly endless debates about scientific data and calls for further demonstrations of proof. All of this moved away from broader, interdisciplinary research questions about alternatives to the industrial model of agricultural production.

The university, sitting at the intersection of many different disciplines, seemed ideally positioned to approach the larger questions about California's agriculture — particularly those about the fossil fuel and chemical intensities of its agricultural industries, the host of negative consequences, and the ability to ascertain whether the people of California would be better served by another kind of agriculture. This implied not simply the basket of techniques, but the production of knowledge necessary to advance a very different model of agriculture. It was in this sense that agroecology seemingly occupied a position deserving of significant resources and institutional support. In the coming years, however, the university would pursue a generalized path of research best described as reductionist. Voices across the university — from the health sciences, schools of medicine, law faculty, social science professors, and many others of whom represented important thought collectives were, instead, left to operate within their disciplinary silos, bereft of a unifying rationale that might have provided for the basis for a revolution in the agricultural sciences. But this alternative mission was not to occur in the coming decades.

Scientific Reductionism and Its Political Consequences

Between the publication of *Silent Spring* and the present day, the application of agricultural chemicals would expand tremendously on a global basis. While regulatory frameworks would grow too, possessing robust information about chemicals and their effects always lagged behind their sale and release, sometimes by many decades, as did understanding of their consequences on human health, terrestrial and aquatic habitat, and larger ecologies. Even as agrichemicals became more regulated than many other industrial substances, workers and their communities would continue to be subjected to multiple exposures, various of which were characterized as having the potential to disrupt human reproduction, interfere with neurological functions, cause chronic diseases, and impact other human biological functions.

Yet even as the evidence against synthetic pesticides increased, these findings typically remained isolated from larger critiques of industrial agriculture. In a certain respect, this isolation followed the design of research pursued by so many agricultural scientists: a reductionist scientific approach, in which findings focused on cellular-level observations were rarely, if ever, linked to higher-level observations, such as the larger ecological, public health, or social consequences of industrial agriculture.

The reductionist approach for addressing pesticide hazards meant that while specific pesticides or techniques might be restricted or changed, the architecture of agricultural production would remain intact. Challenges to this pattern arose among scientists, activists, and practitioners, whose projects operated by linking observations across silos and who pursued a different question: is there a better alternative?

One arena departing from this pattern was groups of scholars and scientists whose work was linked to organized labour. It is here that much cross-pollination occurred among university programs in law and public health. By the early 1980s, legal advocates working with agricultural workers had begun to assemble information from physicians and medical researchers documenting harms from pesticide exposure. Along with filing judicial petitions against growers, the workers' lawyers introduced legislation to identify and restrict those pesticides posing specific harms to workers. Meanwhile, companion legislation was often introduced to provide similar protections to other groups, such as consumers, when similar exposure hazards could be documented in food, water, or the general environment. By the late 1980s, California had enacted a series of laws greatly restricting the use of pesticides in agriculture and other settings, including a prohibition against the sale and use of thousands of pesticide products. The combined force of California's farmworkers' union, their lawyers, public health researchers, environmental and consumer advocates, and other allies shaped a legal strategy influencing laws in various other states as well as federal statutes (e.g., the reauthorization of the Federal Insecticide, Fungicide and Rodenticides Act).

Without dismissing the tangible benefits for farmworkers, there were limitations with legal tactics based on altering technologies. One of the most immediate obstacles arose with nations having very different legal frameworks or a different composition of political forces, especially regarding the power exercised by multinational corporations and national oligarchs versus that of campesinos and others. The stereotype that other nations were at a political disadvantage to alter industrial agriculture was about to be challenged by a group of researchers and workers who were beginning to forge a movement across Latin America: the movement of agroecology. Unlike California's model of legal activism, the agroecological movement advanced a model that was a viable alternative to industrial agriculture.

While the California approach was effective at removing health hazards affecting agricultural workers, their communities, consumers, and the environment by targeting the otherwise unrestrained sale and use of a large assemblage of extremely hazardous chemicals, it was constrained by a legal system steeped in the sanctity of private property and unrestrained commerce. Without dismissing the substantial and significant advances for protecting the health of farmworkers and others, it was an activism that frequently avoided asking a more fundamental question: Who possessed the right to structure agricultural production in the first place?

Notes

1. The scientific group referred to here is the Office of Environmental Health Hazard Assessment, a branch of the California Environmental Protection Agency.
2. Notable scholars included Professors Isao Fujimoto, Don Villarejo, William Friedland, Margaret Fitzsimmons, and many of their colleagues working at different campuses of the University of California.

9 Agrichemicals and the Law

One of the most intriguing ways for US citizens to participate directly in the making of laws is through a procedure known as the initiative process. Emerging in the early 1900s, this process was largely a popular response to curb the overwhelming political influence of wealthy individuals and industries on state legislatures and lawmaking. It enabled voters to act directly by proposing laws to be voted on in elections. Between 1911, when California first adopted this process, and 1989, around 680 initiative measures were presented to voters to create new laws (Jennings 1990: 28–9).

Over the course of its first seventy years, the initiative process proposed new laws on topics ranging from campaign contribution limits, property tax changes, and bond measures to financing new projects. By the 1980s, the growth of citizen-approved initiative measures had expanded at such a pace that by the end of the decade, approximately 40 percent of all the initiative laws introduced between 1911 and 1980 had been enacted in its final ten years, from 1970 to 1980. It was during these latter years that major industries, especially agribusinesses, became increasingly concerned about the progressive content of ballot initiatives and the stricter regulations that could impact their production practices.

While the movement to downsize government was advancing nationally, the political winds in California were blowing in the opposite direction, partly due to a public increasingly displeased with the weakness of federal laws regulating major corporations. As a result, even as President Ronald Reagan undermined federal environmental programs, California's voters pressed legislators to enact even more aggressive state laws — particularly concerning the harms posed by agricultural chemicals.

One of the most dramatic initiatives to be enacted occurred on November 6, 1986, when nearly two-thirds of California's voters approved the Safe Drinking Water and Toxic Enforcement Act, one of the most significant laws to regulate toxic substances in the twentieth century. This law, also known as Proposition 65, prohibited the release of any substance known to cause cancer or reproductive harm into drinking water sources.

Even more upsetting for the industry was the "bounty-hunter" provision, which allowed individual citizens to enforce the law and collect substantial monetary fines from violators if state authorities failed to act. In contrast to vague federal laws that had often failed to protect citizens from criminal corporate behaviour, this provision led to a surge of legal actions by both the state and its citizens against corporate violators, including a number of agribusinesses.

The California ballot measure ignited an intense debate over the adequacy of public health protections provided by the federal government. A range of federal agencies were collecting mounting evidence documenting the inability of federal authorities to identify and control toxic substances, many of which were known to cause cancer and reproductive or other harms to people's health.

At the same time, various industries mounted their opposition to the California initiative. California's agribusinesses were especially vocal in their opposition, arguing that if approved by the state's voters, the consequences would be ruinous for the state's economy. A sample of these messages included the following from agricultural industries or their allies (Jennings 1990: 27):

"A yes vote cast will bring chaos to our state's court system, cripple our economy, and send California's technology and business improvement back some 20 years" (*California Grape and Tree Fruit League*); "Proposition 65, if passed, would have a sweeping and profoundly adverse effect on California agriculture" (*Californians Against Toxics Initiative*).

As aggressive as Proposition 65 was in targeting hundreds of substances known to cause cancer or reproductive damage, a vastly larger number of substances with unknown health effects remained unexamined. While many scientists avoided addressing the larger social and environmental consequences of their work, others offered statements with a different message than the industry warnings forecasting economic ruin. In testimony presented to the US Congress, one courageous scientist spoke about the ever-expanding hazards posed by thousands of synthetic compounds:

> In the early 1940s annual production of synthetic organic chemicals was about one billion pounds. By the 1950s, this had reached 30 billion pounds, and by the 1980s over 400 billion pounds annually. The overwhelming majority of these industrial chemicals has never been adequately, if at all, tested for chronic toxic, carcinogenic, mutagenic and teratogenic effects, let alone for ecological effects, and much of the limited available industrial data is at best suspect. (Epstein 1987: E 3449)

Public concern was quickly turning toward a more comprehensive approach to address the cascading harms surrounding not just groups of chemicals, but the larger process of industrial agriculture.

The opportunity for Californians to fundamentally change state agriculture occurred on November 6, 1990 — a day that its millions of voters were faced with numerous choices that included a proposed law known simply as "Big Green."

Big Green

Big Green, or Proposition 128, was prepared as a statewide initiative for voters to decide on in the November 6, 1990, general election. It promised significant changes to California's laws governing pesticides, food safety, air pollution emissions, old-growth forest preservation, coastal protection, and the enforcement of state environmental laws. Big Green's provisions included the following (Jennings 2017: 33–4):

- Regulation of pesticide use to protect food, agriculture, and worker safety;
- Phasing out of pesticides used on food known to cause cancer or reproductive harm;
- Reduction of greenhouse gases contributing to global warming;
- Imposition of fees on a variety of oil production activities;
- A new elected official to enforce environmental laws;
- Appropriation of $40 million for alternative agricultural research.

The initiative was breathtaking in scope. More than simply an effort to reform existing laws, its authors were forging a path toward a new economy. A principal theme was to eliminate toxic products at the point of production rather than placing a burden on consumers to identify and avoid such products. The greenhouse gas reduction plan was especially striking, calling for a 20 percent reduction in CO_2 emissions between 1988 and 2000, followed by an additional 20 percent reduction between 2000 and 2010.

The defeat of Big Green was largely due to the money that large businesses poured into their opposition campaign. "Over 13 million, 99.5 percent of the opponents' money came from business interests led by three oil companies ... over $8 million came in contributions between $100,000 and $1,000,000. And of those amounts, nearly half of the money originated from sources outside of California" (Jennings 2017: 35).

Big Green's approach was anchored on imposing stricter restrictions and prohibitions over a range of industrial activities. It also provided tens of millions of dollars for alternative agricultural research. However, the initiative's authors struggled to arrive at a more precise definition of what this alternative agriculture would look like. Was it defined by certain technologies, protocols, or procedures? And where did workers and their rights fit into this new agriculture? While the list of prohibited activities, such as the application of damaging synthetic pesticides, could be agreed upon, a default assumption was that free markets and consumer demands would encourage the development of more appropriate, more healthful, and less damaging technologies.

Many of California's decision makers maintained a belief that scientists and entrepreneurs would combine efforts to advance a new agriculture. Yet, as the agricultural research within the University of California had demonstrated, there were already reasons for being skeptical about the nature of free-market ideas and commerce.

A deeper examination of the events surrounding the defeat of Big Green shows that big industry had not simply financed the defeat of a democratic process; it also buried some of the most damning evidence indicating the urgency in shifting away from industrial agriculture.

Hiding the Evidence of Harm

Years later, the role of oil companies in defeating Big Green would expose an even darker side of a covert agenda to misdirect public action regarding climate change (Jennings 2017). The fact that a central actor in this drama — the Exxon Corporation — had earlier been a part of the Rockefeller empire was less important than the reality that it was one of the largest petroleum corporations in the world. More than a decade earlier, in July 1977, Exxon's research and engineering division had delivered a chilling assessment to the company's corporate headquarters: carbon dioxide emissions from global fossil fuel use would warm the planet and could eventually endanger humanity. The same top experts from this division warned again that a doubling of carbon dioxide would increase average global temperatures by 2 to 3 degrees Celsius, and by as much as 10 degrees Celsius at the poles. Rainfall might get heavier in some regions, and other places might turn to desert (Banerjee et al. 2015). Confronting scientific evidence that its core business posed a threat to the planet, Exxon launched a research team to specifically investigate the link between fossil fuel emissions and climate change.

Coinciding with Exxon's investigations into burning fossil fuels and climate change, a larger task force was formed by the American Petroleum Institute (API), enlisting scientists from major oil companies, including Exxon, Texaco, and Shell. In 1979, the task force was briefed by a Stanford University professor whose findings were even more startling than those presented by Exxon in 1977. The presentation began by noting that the primary problems associated with threats to the climate stemmed from the burning of fossil fuels. The threat, moreover, was by no means trivial. His modelling concluded that the amount of CO_2 in the atmosphere would double by 2038, effecting a 2.5-degree Celsius rise in global average temperatures with "major economic consequences." Suggesting even worse to come, "he then told the task force that models showed a 5 degree Celsius rise by 2067 with catastrophic global effects" (Banerjee 2015).

In the following years, the scientific findings reached by the groups of scientists employed by Exxon would not simply be ignored, they would be deliberately hidden from public view. The dramatic conclusions reached by the corporate scientists were only discovered decades later by public interest lawsuits filed against multiple petroleum companies. The discovery would result in a wave of additional lawsuits being filed against Exxon and others, many of which are still pending in US courts.

Alongside findings of corporate fraud that could have altered public policies relating to global climate damages stemming from petroleum interests, many of these same US petroleum firms actively worked to undermine public actions to reduce the use of fossil fuels. Among the most important of these public efforts was a law presented to the voters of California for their approval in the general election of 1990.

The pivot from an impartial scientific forum, as described by investigative journalists, marked the initiation of a project designed to dominate the public sphere.

> Toward the end of the 1980s, Exxon curtailed its CO_2 research. In the decades that followed, Exxon worked instead at the forefront of climate denial. It put its muscle behind efforts to manufacture doubt about the reality of global warming its own scientists had once confirmed. It lobbied to block federal and international action to control greenhouse gas emissions. It erected a vast edifice of misinformation that stands to this day. (Banerjee et al. 2015)

The verified and recorded history shows that one of the world's largest petroleum companies knew its profits came from an industrial process that unleashed global hazards that would persist for centuries.

One cannot know with certainty what the outcome of the 1990 election would have been if the public had been aware of the findings of Exxon's scientists. It is reasonable to believe that California's voters would have approved Big Green, establishing what would have been one of the world's most aggressive plans for moving away from fossil fuels in a state representing one of the world's largest economies. Yet, even if Big Green had passed, changing the trajectory of industrial agriculture required the enactment of a broader array of laws. While some of such laws would be passed in the coming years, the public confronted a constant struggle against the insistence by agribusinesses and agrichemical firms that California adhere to a model of industrial agriculture.

A fascinating aspect of this political drama is that in the aftermath of the defeat of Big Green, a group of Latin American scientists had already arrived at the threshold of defining an alternative to this industrial model.

Part III

The Struggles of a New Science

More recently, agroecologists have shifted their attention from a critique of industrial agriculture to demonstrating the value of agroecology as a viable alternative, establishing a bridge between ancient agricultural systems and a science anchored in a set of social principles. During this period, agroecology has been tested both in terms of gaining recognition as a legitimate science as well as responding to the rising political and economic conflicts affecting campesinos, rural households, and social groups across the Americas. Even as agroecologists have expanded their influence in many social arenas, they have also experienced strategic setbacks, including their struggle to gain acceptance in university settings.

It is particularly the story of agroecology's auspicious beginnings at the University of California, Berkeley campus that raises a number of troubling questions relating to contemporary crises of scientists pursuing a new science: "How did agroecologists simultaneously advance a viable alternative to industrial agriculture, support embattled campesinos across the Americas, and defend their work in a hostile university environment?

10 An Alternative in Production

I would later reflect on my introduction to agroecology and the expanse of my ignorance when I first journeyed across Mexican landscapes. Indeed, my companion during many of these journeys, Cheri Lucas Jennings, reminded me how I nearly dismissed my very first tutorial about alternatives to industrial agriculture. The tutorial began with a generous offer by one of Miguel Altieri's graduate students, Javier Trujillo Arriaga, who introduced me to his colleagues around Chapingo.

Javier offered us a visit to the floating gardens or chinampas in Xochimilco, one of the older sections of Mexico City. These chinampas are the remnants of a vast canal system developed hundreds of years ago by the Aztecs in Tenochtitlán, a region now largely occupied by Mexico City. Prior to the arrival of Spaniards in the sixteenth century, chinampas had been constructed on an estimated 25,000 to 38,000 acres (Wright 1991: 158). Cheri recalled that I was almost rejected the offer, thinking Javier was simply looking for a way to entertain his guest. My impression of a classic tourist trap was only reinforced at the first sight of a flotilla of gaily festooned and colourful boats. Faced with Javier's obvious enthusiasm, I put aside my reluctance and boarded the small boat at the entrance to an extensive series of canals.

It was only as they made their way through the intricate channels that Javier explained the genius and complexity of this agroecological system. He pointed out the selection and placement of different plants, their intricate and carefully structured relationship in terms of allocating light for photosynthesis, the contributions and sharing of soil nutrients, the recycling of carbon from the channels, the control of pests and diseases offered by various plant mixtures, and their plentiful yields for an array of purposes: food, medicine, textiles, and more. The scientific sophistication of these thirteenth-century marvels of ecological engineering lay in their integrated design: a system for regulating water flow, the intensive cultivation of elevated lands through complex agroecological practices, and the simultaneous transport of abundant foods to central markets (Mundy 2018). Once introduced to such an example of agroecology, it was difficult to unsee. The same sentiment applies once one comes to recognize the role of campesinos in developing and managing their agricultural systems.

Javier's tutorial on agroecology was later followed by additional discussions about his intellectual journeys in both the United States and Mexico. A central feature of his journey was a dawning realization about the political nature of his education and how the legacy of the Rockefeller Foundation's program in the 1940s continued to reverberate years later. Reflecting on his time at Chapingo in the 1970s, Javier observed that the scientific agriculture model introduced decades earlier continued to marginalize broader forms of knowledge, especially those that could reveal the failures of conventional agriculture.

> When I went to Chapingo I noticed that in spite of the enormous prestige of the place there were seriously flawed approaches to pest management problems. The main idea was to study what was done in the United States, studying textbooks in English. We received solutions, that is, we would be given the name of an insect and the four chemicals which had at some time and place been used successfully against it.
>
> Fortunately, it was necessary for me to wait a year before I could enter Chapingo. I used this year to study biology. I had thought of agronomists as primarily biologists, so I prepared by reading and studying biology. With biology fresh in my mind I entered Chapingo and was immediately conscious of something wrong there. There was a religious sense in Chapingo — what was taught in Chapingo you couldn't question. In addition, there was a constant fight for power at Chapingo that tends to preoccupy people there, because what happens at Chapingo has such enormous national consequences. But the basic problem for me was that at Chapingo there is very little or no ecological or biological preparation. At the postgraduate level you find some consciousness of the biological and ecological problems, but the people at the postgraduate level have almost nothing to do with the undergraduates. Thus, you have hundreds of agronomists every year given their degrees without any real consciousness of what they are involved in. Those who do know better are not prepared to enter into the struggle for power at Chapingo, and so things remain the same. Chapingo remains the guide for the whole country. (Wright 1991: 66–7)

It is worth noting that Javier, like many others at Chapingo, occupied a political terrain rife with conflicts and contradictions. On the one hand, alternative agriculture presented an appealing moral option — advocating a more socially just, ecologically sensitive, and caring form of agriculture for consumers and for campesinos and their families. It sought, in short, to build an agriculture grounded in a higher moral imperative. Others at Chapingo, however, understood Mexico's agricultural problems as inseparable from its capitalist economy; for them, resolving these conflicts required far deeper social and political change. This tension — between the moral appeal of alternative agriculture and the structural analysis advanced by many agroecologists — would become a crucial divide in the years ahead. In the meantime, many young scientists found themselves in unfamiliar terrain, trying to situate their work within a landscape of political choices.

Even so, Javier found avenues for raising more than just questions about the nature of education that the Foundation had fostered at Chapingo. Following the completion of his doctorate at the University of California at Berkeley, Javier soon assumed teaching responsibilities at Chapingo and the nearby graduate school at Montecillo. His initial instruction aligned with the traditions established by the Rockefeller scientists decades earlier: students were introduced to the same kinds of techniques, machinery, and modern scientific methods as would be on display at agricultural colleges in the US. However, Javier typically presented one fundamental twist: during their final exams, he would require his students not only to demonstrate knowledge of modern agricultural approaches but also to critically assess their limitations, particularly in relation to the broader ecological context. Thus, students experiencing Javier's instruction would not only learn about increased crop yields but also critiques of consequences ranging from techniques that might foster animal or plant diseases to plant regimens depleting soil fertility or contaminating groundwater. Javier's pedagogy reflected that of a small but growing band of academics across the Americas who — having learned from such scholars as Xolocotzi, Altieri, and others — became active participants in establishing a new generation of agroecologists.

In a meeting at the graduate school in mid-1991, more than a dozen researchers presented their findings on irrigation, forestry, natural resources, soils, and livestock. Except for being delivered in Spanish, their presentations largely mirrored the same kinds of research taking place at agricultural colleges across the United States.

Javier, increasingly frustrated with colleagues working only within the boundaries of a Green Revolution mentality, challenged them to expand their investigations beyond the boundaries of what had been shaped by

An Alternative in Production 91

the Rockefeller scientists. "Why," asked Javier, "if the mission of Mexico's agricultural institutions has become truly scientific, is there not also an investigation into other methods of pest control?" He then launched into his larger presentation on an increasingly popular theme: Why are alternative research paths not being pursued by university researchers?

Javier then presented slides showing that over the previous twenty years (1970–1990), roughly 50 percent of all pest management investigations were based on testing the efficacy of pesticides, whereas other techniques, including biological controls, accounted for only 5 percent. Briefly summarizing the broadening array of pesticide-related problems, Dr. Trujillo then asked his colleagues about the curious bias in their research design: "If the project here were simply one of science, how is it that 50% of the investigations favoured pesticide-based controls?" In closing, Javier posed a question not frequently articulated at such gatherings: "Research for whom?"[1]

Javier's primary research interest, however, was not to spend much time critiquing the failures of synthetic chemicals in agriculture; rather, he was eager to bolster a more exciting alternative to the model that the Rockefeller team had delivered to Mexico decades earlier.

Returning with Javier from his presentation at the graduate school, we stopped at his office where he displayed clippings from regularly collected newspaper articles on issues affecting agriculture across Mexico. It was readily apparent why Javier was preoccupied with researchers' favouring pesticides as a primary tool; here is a sample of articles reported in two of Mexico City's major newspapers during 1991:[2]

March 23: According to the president of the Independent Federation of Agricultural Workers and Peasants (FIOAC), as a result of the 36,000 tons of pesticides applied each season in Sinaloa, there is a high rate of death among children who work the fields as contract labour in tomato, pepper, melon, and other export crops. *Seguro Social* reports that the average of fifty deaths among children during each harvest in the region is likely an underestimate of actual fatalities.

March 27: The district chief for rural development in Mazatlan, Sinaloa, complains that the indiscriminate use of pesticides is contaminating lagoons, rivers, and estuaries.

May 23: The World Health Organization reports that La Comarca Lagunera ranks among the highest in world surveys for incidence of degenerative diseases such as cancer. The WHO identifies the use of extremely hazardous pesticides in the region as one of the most probable factors.

June 5: A professor with Universidad Autonoma de Yucatan claims that thirty-five pesticides prohibited in the United States are regularly being used in the region on various crops.

July 22: A study by la Escuela de Ciencias del Mar de la Universidad Autónoma de Sinaloa identifies pesticides entering bays and estuaries as causing a 50 percent decrease in the catch of camaron, robalo, and other species between 1979 and 1989.

July 26: Aerial applications of pesticides in Sonora, according to studies conducted by SEDUE [a Mexican federal agency with responsibilities for environmental policy] threaten the health of fifty thousand families in the Yaqui and Mayo Valleys.

October 7: A spokesperson for the IMSS states that there are four to five cases of childhood leukemia deaths every month in the Ciudad Obregon region and that in recent years leukemia deaths have risen by 30 percent for children and 70 percent for adults. Aerial applications in nearby cotton fields are a suspected cause of the disease.

October 20: The president of the Strawberry Association in Irapuato confirms that during the last season only 20 percent of strawberry exports to the United States were accepted, with much of the 2,250 tons harvested denied entry due to the detection of illegal residues.

October 22: The secretary general of La Central Independiente de Obreros Agrícolas y Campesinos (CIOAC) says that large agribusiness operations in the Valley of Culiacan, in the south and coastal regions of Sonora and Baja California, apply pesticides from the air while workers are in the field. Medical findings show that workers have suffered damage to sexual organs, including infertility, in both men and women.

But Javier and his colleagues had already moved beyond the recurring reports on the failures of the Green Revolution, turning their attention to the realm of alternatives.

Producing an Alternative

In the fall of 1991, posters appeared across the campus at Chapingo announcing a symposium entitled "An Alternative in Production." Presenters included Efraím Hernández Xolocotzi, Miguel Altieri, Steve Gliessman, Julia Carabias, and myself, among a host of others. While many were familiar with the works of the first three scholars, Dr. Julia Carabias was a relatively new and impressive addition to the growing numbers of faculty drawn to a topic garnering increasing attention, especially among a younger academic audience.

Dr. Carabias's presentation was especially striking as she explained the situation Mexico now confronted as a result of the Green Revolution. The traditional practices that for centuries had provided a bountiful harvest of foods, fibres, medicines, forest products, fish, and animals were disappearing. In their place was a package of modern techniques that resulted in the massive destruction of soils, contamination of vast bodies of drinking water and aquatic habitats, poisoning of uncounted numbers of workers, and low-level warfare conducted against landless campesinos (Carabias et al. 1989).

Many members of her audience at Chapingo were already familiar with a study she had authored with one of Mexico's most well-known ecologists, Victor M. Toledo. In that work she and Toledo documented and argued that the Green Revolution was obliterating more than five thousand years of accumulated knowledge; a knowledge that formed the basis for agroecological practices tailored to a wide variety of Mexico's diverse landscapes (e.g., marceno, campos elevados, huertos familiares, cacatales, and chinampas in the warm, humid tropics; terrazas y policultivos in the sub-humid temperate or templada; campos drenados, cafetales bajo sombra in the humid temperate zones; cultivos de escorrientia in the arid and semi-arid regions; and policultivos sobre dunas, pasto marino en la costa) (Carabias and Toledo 1983: 26–7).

In the year before Carabias's presentation at Chapingo, another of her colleagues had noted the changes being wrought on Mexico's food landscape by tracking the impact of commercialized agriculture on traditional agriculturalists in Morelos since the 1970s. In looking at the small village in Morelos, the investigator had noted that the introduction of monocultures of avocado and peach production displaced an intricate mixture of many other food crops, including twenty-eight different polycultures based on such foods as peach, pear, lime, calabaza, avena, maize, and other crops. This resulted in the gradual elimination of an entire system of family gardens with their origins in mesoamerican traditions: a system based on local germplasm, traditional knowledge of how to construct and maintain these polycultures, as well as serving the needs of local and regional markets (Ortencia 1990).

In the years following that first symposium at Chapingo, researchers would gather many more scientific studies explaining the genius of so many traditional agroecological systems across the Americas. These very specific scientific case studies would be the first steps in demonstrating the viability of agroecological approaches for food production. Even more important, such studies opened the door on other metrics, other standards, and the fostering of a very different set of questions to guide what was becoming a distinct body of knowledge. Agroecologists were rapidly pursuing what other philosophers of science would abstractly refer to as a scientific revolution.

Notes

1. The preceding section is based on a symposium entitled "Scientific Investigations in the Agricultural Sciences in Mexico" and attended by the author at the postgraduate school, Montecillo, on August 16, 1991.
2. The newspaper accounts are contained in articles appearing during 1991 in two Mexican newspapers, *Excelsior* and *El Universal*, and collected by the Department of Parasitology at UACH (without the benefit of further citations).

11 Demonstrating Another Knowledge and Practice

Professor Steve Gliessman, another of the speakers at Chapingo's first agroecology symposium, also joined with various of his colleagues in celebrating an alternative to the Rockefeller scientists' push for industrial agriculture.[1] In a book that Gliessman authored the year before the Chapingo conference, he praised Miguel Altieri's work that dovetailed with the theme of the symposium — identifying an alternative in agricultural production:

> Rather than dwelling so heavily on the problems of modern conventional agriculture, his [Altieri's] book went much further in describing a theoretical foundation for the study of agricultural ecology by presenting examples of agroecosystems that incorporate the concepts of ecology into their design and management. His examples ranged from traditional Third World agroecosystems to small-scale alternative and organic systems in developed countries. (Gliessman 1990)

One had the distinct sense of new excitement among Chapingo's students. The very fact that the symposium's speakers guided their audience to treat agroecology as not simply a new set of techniques but a complex scientific undertaking that blended food production, restoring local ecologies, and the rights of campesinos was something not previously experienced at Chapingo. It was a scientific revolution. The audience reception following each speaker was frequently thunderous appreciation. It was especially the enthusiasm of students that served as an indicator of broader support. The 1991 symposium on agroecology at Chapingo was quickly followed by two more international symposia on agroecology in 1992 and 1994. In a very real sense, the early 1990s marked a period of rapid expansion of agroecology at universities across Mexico (Astier et al. 2017).

Later in 1991, a group of professors at Chapingo created a program in agroecological engineering.

Located in the center of the country, this university program is influential at the national level, with the majority of its graduates and projects located in the central, southern, and southeastern regions of the country. It aims to train agroecologists who are capable of proposing solutions to environmental problems stemming from conventional agriculture and its impacts on rural life, with a focus on interdisciplinary training. Teaching practices have been developed over the last 20 years through processes of participatory management with farming communities and with a multidisciplinary focus to overcome the fragmentation of knowledge. There are now 535 graduates of the program, the first having earned their degrees in 1995. (Astier et al. 2017)

The Colegio de la Frontera Sur (also known as ECOSUR), established in 1994, began offering various courses, and then master's and doctorate degrees in agroecology, sustainability, and pest management in 1995. As many of its faculty members had studied under Xolocotzi, the curricula reflected a broad perspective, drawing from a group of twenty-four researchers with expertise in germplasm, soil ecology, landscape ecology, farming territories, traditional knowledge, sustainable food systems, and social movements. Indeed, in many important respects, the centre for Mexico's university-based work on agroecology shifted to ECOSUR and its very full complement of faculty who had not just advanced university training but were deeply familiar with campesinos and their efforts to organize politically.

By the mid-1990s a number of other universities across Mexico had initiated courses in a variety of disciplines relating to agroecology. The rapid emergence of agroecological programs and courses across the country, involving dozens of faculty training many dozens of students annually, mirrored what was occurring across many parts of Latin America and led to the spread of investigations, conferences, and publications on the topic. The enormous interest and discussions at various universities resulted in the launching of innumerable research projects across Mexico and the Americas, and these helped move agroecology beyond the boundaries of the university and into the realm of applied work. Perhaps most crucially, these agroecological projects frequently operated with the active participation and direction of campesinos.

Paralleling the meteoric rise of agroecology among students, campesinos, and many others, the IARCs were awakening to the emergence of a scientific revolution on their doorstep.

Fixing Industrial Agriculture

Even before institutional organizing of agroecology began to blossom across the Americas, the Rockefeller and Ford Foundations, along with other supporters of industrial agriculture, were aware of the dysfunctional nature of its "successes" on the ground. Much of this surfaced with the expanding critiques of the Green Revolution, especially with regard to its multidimensional negative consequences involving displacement of campesinos and their communities, income inequalities, pollution and environmental damages, worsening diets, dangerous workplaces, and so on. The response by the foundations, international donors, and agribusinesses involved a cascade of "fixes" designed to repair what were generally regarded as unintended consequences of an ever-evolving technology.

Among the earliest generation of IARCs, such as the CIMMYT and IRRI, the integration of economists into the work of agricultural technologies was intended to provide national governments with avenues for adjusting programs via policies. These policies provided a distancing of the responsibilities that Green Revolution scientists might otherwise face regarding the larger consequences of their work. "At CIMMYT one sign of the shifting terrain was a manual instructing agricultural researchers in the development of technologies suited to farmers' needs … with the publishing of a booklet by members of CIMMYT's Economics Program: Planning Technologies Appropriate to Farmers" (Curry 2022: 208).

The Rockefeller Foundation attempted to jumpstart this kind of research within the still-growing CGIAR network in 1974 by offering postdoctoral fellowships at the IARCs to social scientists. These "farmer-centred" approaches to agricultural development from the mid-1970s onward went by different names, such as Farming Systems Research, Farmer-Back-to-Farmer, and Farmer First, but shared in common social science methods and the desire to incorporate farmers' perspectives into research strategies and technology development (Curry 2022: 208).

Yet the distancing that the IARCs sought to create was quickly revealed to many local observers as a thinly disguised attempt to avoid responsibility for what were inherent design flaws. Independent examinations of these pilot programs for creating a second generation of Green Revolution projects, like CIMMYT's Puebla Program, underscored the fact that central features of industrial agriculture were constructed to meet the exigencies of capital formation and extraction and its integration into international markets and free trade. The treatment of workers, their communities, regional and global ecologies, and democratic control over an economy were, at best, adjustments subordinate to the paradigm of production. As my earliest collaborator, Dr. Edmund K. Oasa, was fond of saying, the new generation of farmer-centred programs offered by IRRI and CIMMYT to address the damaging effects of their technologies were simply "old wine in new bottles" (Oasa 1981a).

The rapid growth of agroecological field demonstrations, rural schools, university programs, and regional conferences were soon weaving together across the Americas multiple networks of people advancing the process. These new networks mirrored in certain respects the early approach adopted by the US agricultural scientists — except for the fact that the process of spreading agroecology inverted the central features of the Green Revolution scientists. First, agroecology began its work by advancing a campesino-based agriculture for those at the bottom, not the top, of the economic pyramid. Second, its work centred on using traditional forms of agricultural practices, practices documented as delivering stable yields that maintained or improved the surrounding ecology. Third, it fostered first and foremost a viable household economy for those working the land, providing a foundation for supporting thriving regional economies. Finally, like the Green Revolution scientists, the work of agroecologists produced another set of practices and knowledge. However, there was a singular difference: in place of a scientific revolution countering the just social order that millions of campesinos had fought to create for those working the land, agroecologists specifically sought to restore the democratic principles for advancing the political rights and social justice for the largest class of workers, the unemployed, and the marginalized members of society. This notion of justice included access to healthy and nutritious diets.

A Gathering of Other Scientists

Establishing a network of people working on agroecology throughout the Americas began with the creation of the Latin American Consortium on Agroecology and Development (CLADES — Consorcio Latinoamericano sobre Agroecologia y Desarrollo) in 1990, with Miguel Altieri as a cofounder. Convened by a group of eleven non-governmental organizations and representing eight South American nations, CLADES functioned as an institutional base to promote investigation, training, and information about agroecology. Its researchers demonstrated repeatedly that agroecologically managed systems provided stable levels of production, economically favourable rates of return, enhanced biodiversity and soil health, as well as a stable livelihood for small farmers and their families (Altieri and Nicholls 2017).

In its early years, CLADES's multiplicity of tasks included the convening of conferences, the publication of ongoing agroecological field research, and conducting courses across the Americas. It sponsored annual gatherings that initially drew dozens of individuals, then a flood of hundreds, and then thousands of enthusiastic participants (Sanchez 1992). The intellectual seeds it planted blossomed with the creation of a network of agroecologists extending from local groups of campesinos to university researchers and a host of allied organizations.

Another of the organizational homes for agroecology appeared with the establishment of a scientific society in 2007 — the Latin American Scientific Society of Agroecology, also known by its Spanish acronym, SOCLA (La Sociedad Científica Latinoamericana de Agroecología). For the initially small group of scientists spread across the Americas, SOCLA served as a vital gathering place to discuss both the substance and purpose of agroecology. It was another organization in which Dr. Altieri served as a founding member.

SOCLA worked to identify and define the multifaceted characteristics of agroecological systems (Altieri and Nicholls 2017).

SOCLA's achievements included the creation of two doctoral programs in agroecology at major universities in Spain and Colombia, the sponsoring of numerous short courses in various parts of Latin America, as well as organizing regional research programs and conferences (Altieri and Nicholls 2017).

Whereas many other scientific societies were dedicated to insular pursuits focused on discussions among their own scientists, CLADES and SOCLA purposefully orchestrated meetings and conferences designed to address broader issues beyond the usual scientific boundaries. To this end, they sought to link scientists and nonscientists, especially campesinos, who became full participants in defining this emerging science. CLADES and SOCLA in this way served as a forum for introducing democratic participation into the fabric of devising another realm of knowledge about the theory and practice of agroecology.

This knowledge increasingly joined with knowledge from other disciplines, such as ethnobotany, public health, and alternative economics. It was precisely this characteristic knowledge expansion that helped agroecology to produce what have been elsewhere termed as epistemological innovations (Toledo and Barrera-Bassols 2017). Agroecology was becoming a science merged with applied knowledge, designed to overcome the kind of wilful ignorance practised by Green Revolution scientists.

In their merging of scientific and social purposes, CLADES and SOCLA were not simply just another group of scientists solely focused on increasingly finer details. Their research, presentations of findings, and discussions constantly moved between field observations of biological events and the understanding and assessment by campesinos and their communities.

CLADES and SOCLA fostered the emergence of agroecology as not just another variant of industrial agriculture, but as a new discipline, a revolutionary science.

Note

1   The first pages of this chapter draw on the publication authored by a group of twenty-three scholars (Astier et al. 2017), largely at Mexican universities, describing the development of agroecology in Mexico. See <pmcarbono.org/pmc/publicaciones/Back_to_the_roots_understanding_current_agroecological_movement_science_and_practice_in_Mexico.pdf>.

12) New Markets / New Conflicts

Following the discussions at Chapingo on production alternatives in early March, 1990, a group of Chapingo's students and faculty had already arranged for me to give a separate presentation at the CIMMYT. The presentation encapsulated many of the same critiques contained in my doctoral dissertation, and these also appear in the early chapters of this book.[1] Even before completing my brief presentation at the CIMMYT's indoor amphitheater, the gathering of CIMMYT scientists and staff erupted with angry reactions to my criticisms of their work. After they had stood and shouted their objections, I invited their questions.

As much as the audience at Chapingo had been celebratory with anticipating a new path for research, the CIMMYT's audience, composed largely of its professional staff, were hostile and defensive. Their objections centred principally on two points: first, the history described by CIMMYT critics misrepresented their achievements in Mexico and across the globe; and second, the CIMMYT had already embraced many of the earlier criticisms and changed its ways. However, unresolved was a crucial question: To what extent were the CIMMYT's reforms merely ones of putting old wine in new bottles? How would its reforms alter the exploitive political and economic context that it had helped to create many years earlier? In closing my presentation, I offered that if indeed CIMMYT had changed its ways, surely their scientists could initiate a much broader public discussion of its priorities. Inspired by the just-concluded conference at Chapingo, I stated that a useful place to begin this public discussion should begin with the CIMMYT's pursuit of various biotechnologies. My recommendations for a much broader public discussion surrounding the scientific work of the IARCs would re-emerge in others' discussions in subsequent years.

The outrage expressed by the CIMMYT professionals was received with a mixture of shock and dismay by many of Chapingo's faculty and students, who had travelled the short distance to witness the exchange. Departing from the CIMMYT, various students and professors from Chapingo wondered aloud if anyone at the CIMMYT had even attended the earlier symposium just up the road at Chapingo or understood anything about the many important presentations on agroecology. The reaction by CIMMYT's staff was puzzling. How did one reconcile these starkly contrasting perspectives on what was happening in Mexico's landscapes?

In retrospect, one group later provided a stark counterpoint to the perspective of the CIMMYT scientists. The group of Mexican agroecologists at ECOSUR would argue that the emergence of the distinctive science of agroecology was integrally tied to the social and political struggles occurring throughout the Americas. As Peter Rosset, who would become a leading voice representing agroecologists across the Americas, and his colleagues would later write, the underlying social and economic conditions of their agroecology were predicated on a set of facts that industrial agricultural scientists were constantly excluding from their work.

> One paradigm — that of development — emerges from Western epistemologies, focusing on the creation of common markets, cartesian science and the increasingly prominent role of the [Latin American] region for commercial and financial capital, in conjunction with reviving the traditional role as provider of raw materials and consumer of manufactured goods.... Meanwhile, another paradigm — the "critical" paradigm — is highly skeptical of Western development discourse, and instead sheds light upon the epistemological diversity of the "many worlds" within Latin America....This second paradigm is part and parcel of long-standing processes of Latin American integration — and the construction of regionalisms "from below," based on movements and accompanying intellectuals from both inside and outside academia who are critical of "the system" and of hegemonic forms of thought. (Rosset et al. n.d.: 635–52)

Following the conference at Chapingo, I was persuaded that rather than engaging in futile debates with scientists steeped in a model of industrial agriculture, a better use of my time would be spent on understanding what was occurring in the landscapes beyond the CIMMYT's well-groomed experimental fields. Travelling by bus from Mexico City to the Yucatan, I then continued on to Chiapas to complete another phase of my research. This phase focused on the US-proposed trade deal with Canada and Mexico, a project generally known by either its English-language acronym: NAFTA (the North American Free Trade Agreement) or as el Tratado de Libre Comercio and its Spanish acronym: TLC.

For months the Mexico City newspapers and academic discussions had been filled with analyses of NAFTA that largely conflicted with projections provided by US negotiators — particularly the official statements that all parties to the agreement would realize considerable economic benefits. As the date for representatives from the three nations to meet in Mexico City drew near, a main focus of many conversations focused on the consequences for Mexico's agriculture. Prominent in the reporting from Mexico City were the many academic sources projecting a cataclysmic impact on Mexico's campesinos.

Among the less-reported facets of the meetings in Mexico City were the uninvited participants and observers. Most striking, perhaps, were Canadian farmers who arrived and stayed in Mexico City for an extensive period to join with campesinos protesting the supposed benefits of the TLC for farmers. During this period, the Canadian National Farmers Union participated in a three-day conference entitled "Agriculture, the Environment and the Free Trade Agreement" in Mexico City. The conference was among the first of many more future occasions when farmers, peasants, and their allies would gather to share information about tactics and strategies to understand the detrimental effects of trade policies on their communities (Desmarais 2007: 81).

While much attention was focused on the remarks by differing experts from the three capitals, Cheri knew that my curiosity was drawn toward the countryside. It surprised her little when she received the call from me that instead of returning immediately to Chapingo, I was going to take a less direct route, travelling instead to Chiapas. "Are you going to stay long in Chiapas?" Cheri asked. "No," I answered. "Only long enough to see if I can have a conversation with the Archbishop."

Cheri knew from news coverage in Mexico City of the NAFTA negotiations that one voice in particular stood apart from that of many economists and trade policy experts; an otherwise little known archbishop serving in a distant town in Chiapas, a place called San Cristóbal de Las Casas. The church leader, Archbishop Samuel Ruiz, was notable for many reasons, but especially for his long-standing commitment to serve the rural poor. While other, more conservative members of the Catholic Church distanced themselves from the everyday livelihoods of Mexico's poor, Archbishop Ruiz was among a progressive group of priests who were deeply immersed in the world of Mexico's campesinos and their struggles.

Days later I stood outside the large cathedral that dominated the plaza of San Cristóbal. I had joined the long line of people, mainly campesinos, who awaited the opportunity to speak with the Archbishop. Standing among others, many of whom were experiencing obvious grief and suffering evident hardships, I doubted my wisdom of thinking this man was going to take the time to converse with me in this place about international trade.

I finally took my turn to ask my question: "Can you please share your thoughts regarding the proposed trade agreement?" His pause made me reflect again about my own audacity to appear before one of Mexico's most highly respected figures and expect a serious response.

The Archbishop's response demonstrated that matters of international finance were deeply consequential for the people of Chiapas. Speaking from historical context, he rejected the premise that the proposed treaty promise of "free trade" delivered something new: "Chiapas and Mexico have suffered the consequences of free trade for more than 500 years." Referring to recent events in Chiapas, he described how a group of hacendados had deployed men carrying weapons in helicopters and trucks to dislodge campesinos who occupied ancestral lands not far from San Cristóbal. The political conflict over the rights of those who worked these lands versus an elite who claimed ownership that had produced Mexico's revolution early in the twentieth century was still an open and bloody conflict in many parts of Mexico.

Explaining that "el tratado de libre comercio" would only worsen the conflicts over the control of land and the exercise of democratic rights, the Archbishop concluded his remarks with a question of his own, "What is there in the nature of the free trade agreement that will change the problems faced by people in Chiapas or elsewhere in Mexico?" This central question for the Archbishop and so many of his parishioners was seemingly lost to a group of negotiators scripting a plan for international trade.

Yet, even those experts who dealt with the technical language of the Tratado understood the potentially devastating consequences for Mexico's campesinos:

> Experts diverged widely in their predictions of NAFTA's effects in the years leading up to its implementation. Advocates argued that NAFTA would increase trade, create jobs, and reduce the cost of food and manufactured products.... The pros and cons for agricultural production loomed particularly large in debates over the desirability of free trade. US grain producers, with the advantages of a temperate climate, state-funded research, and access to credit, insurance, the latest technologies, and extravagant subsidies, outproduced Mexican farmers by huge margins. The US government was desperate to unload cheap grain into Mexican markets. In Mexico officials promoting NAFTA touted the anticipated flood of US grain as a boon for consumers who would see food prices fall, but many predicted the same influx would deliver a fatal blow to Mexico's poorest farmers. (Curry 2022: 212)

It's important to recognize that the widespread opposition to NAFTA was not simply a matter of sentimental or moral outrage. As Archbishop Ruiz and many of his fellow adherents of liberation theology could attest, the advance of capitalist agriculture in Chiapas and beyond was inevitably accompanied by what many economists blandly described as a process of modernization — one that entailed the creative destruction of traditional ways of life. The resistance expressed by so many was, at its core, a political challenge to the logic of industrial agriculture and its entanglement with so-called free markets.

In the years immediately following the formal discussions between the three countries in 1990, the Tratado would be largely forgotten, merely another bookmark to the seemingly inescapable advance of industrialized agriculture under the larger architecture of free-market trade policies. Until one morning, several years later...

Agricultural Technologies, Free Trade, and Structural Violence

On the morning of January 1, 1994, as the Tratado went into force as the new law governing international trade between Mexico, Canada, and the United States, the world awoke to discover that the deceptively quiet and remote town of San Cristóbal de las Casas in the Mexican state of Chiapas had suddenly become the centre of revolution. The Zapatistas, now recognized as a movement initiated by the largely Indigenous people living across Chiapas, sought to explicitly challenge the notion that free markets represented any kind of advancement for Mexico's campesinos and others. The campesinos of Chiapas erupted as a new voice: the Zapatista insurrection. "One of the principal demands of the campesinos was for protection against NAFTA's expected impact on agriculture. For the Zapatistas as well as many others, NAFTA's provisions had the potential to devastate close to two millions Mexican peasants who produced corn, the country's food staple" (Gonzalez 2022: 316), especially those millions of campesinos on small plots of lands.

The impact of the Zapatista movement would reverberate for years to come across the Americas and beyond. For many of Latin America's agroecologists, the Indigenous uprising served as yet another reminder of both the irrationality of industrial agriculture's linkage to global markets and the compelling case for an agriculture grounded in grassroots autonomy and self-government, as advanced by the Zapatistas. It is little wonder that campesino-based organizations, such La Vía Campesina (LVC), found inspiration in the words and political organizing of the Zapatistas. Among the most important recent books on the movement is *Lessons from the*

Zapatistas by Lia Pinheiro Barbosa and Peter Rosset (2025).

Before long, observers noted that the new terms of trade were worsening the position of Mexico's campesinos. "The labor picture became even more dismal once you factored in NAFTA's impact on the Mexican countryside. With government subsidies for sowing corn eliminated, small farmers simply could not compete with the mechanized output of US agribusiness. Mexico's grain imports from the United States tripled from 1994 levels and now represent 40 percent of that country's food needs. Agricultural employment tumbled by nearly 20 percent between 1991 and 2007, from 10.6 million to 8.6 million. Many of those 2 million unemployed peasants were forced to either join the ranks of the country's huge informal economy or migrate to the United States (Gonzalez 2022: 320).

> So instead of slowing down the exodus to the United States, NAFTA, with its extraordinary impact on Mexican agriculture, has sped it up. The Mexican-born population of the United States went from 4.5 million in 1990 to 9 million in 2000, and then to 12.7 million in 2008, with more than half of that population being undocumented. Rural dwellers represented 44 percent of those migrants even though only one-quarter of Mexico's people reside in the countryside. (Gonzalez 2022: 320)

Selective histories treat vanishing campesino communities in the Americas as the result of a march toward modernization or what some economists referred to as a process of creative destruction. Yet, as the Foundation's missions had gained momentum in rural areas across the Americas, the disruptive force of so-called modernization assumed a more active form — what Latin American social scientists have termed "structural violence."

As Green Revolution policies spread across the Americas in the 1950s to 1970s, the violent suppression of peasant protests would surface again and again. The protests by peasant communities would result in even more active efforts designed to forcibly incorporate rural communities into the markets dominated by larger economic interests. The new struggles of the late twentieth century and early twenty-first century would subject even more campesinos to what might more accurately be termed "the killing fields."

The longer-standing struggles in Guatemala, Nicaragua, El Salvador, Honduras, Brazil, Chile, Colombia, Peru, and many parts of Mexico reached a new level with the Zapatistas and the renewed revolutionary struggle regarding who held rights, access to, and control of lands, as well as resources, capital, and political institutions. A central argument for campesinos, urban workers, and the mass of unemployed persons was that liberalized trade and neoliberal economic politics would only further heighten social, political, and economic inequalities across the Americas. Within this argument, the notion that industrial agriculture represented an agreed-upon goal for all humankind faced increasing evidence of the opposite: the Green Revolution and its reformed versions based on free and open markets were hugely destructive, especially for small-scale farming systems.

The historical record of US interventions across Latin America during the twentieth century was predicated on regarding tens of millions of campesinos as an oppositional force to the Rockefeller Foundation's campaign of "starting at the top." Campesinos protesting this treatment — the taking of traditional landholdings, enslavement and inhumane working conditions, the theft of water and natural resources, torture and massacres — were interpreted as threats to the free-market ideology of free trade and private rights. By the 1960s, US policies of a Cold War against communism gave tacit and explicit support to the financial elite, dictators, and right-wing military officers to engage in the active torture, repression, and murder of campesinos — particularly those who threatened the stability of an industrialized agriculture serving US markets (Blitzer 2024).

International Agricultural Research Centres and Agroecology

It is in this context that industrial agriculture and the scientists who designed its technology and practices were increasingly viewed as deeply committed to a particular social-political order — one largely constructed on the subjugation of campesinos, their communities, and their lands. The expansion of agroecology no longer represented simply an effort to document empirical evidence supporting traditional over industrial agricultural practices; it was now merging with political movements advanced by Indigenous and campesino communities to support their land, lives, and livelihoods based on historical agriculture.

The CGIAR, as the umbrella organization for the international agricultural research centres, thus found itself negotiating a rising tide of rural conflicts. Faced with such challenges, in 1995 the CGIAR decided to form an advisory group drawn from NGOs to recommend a path that might extricate the individual centres, such as the CIMMYT, from further controversy.

Following the lead of other international organizations, the consultative group decided it needed to select a recognized agroecologist, preferably from a major university, to lead it. The CGIAR was compelled by numerous agroecologists to select Miguel Altieri. As the chair of the Non-Governmental Committee (NGOC), Miguel and his colleagues quickly assembled a report that turned the tables on the typical use of NGOs. In too many instances, NGOs were massaged with stipends, flights to international meetings, and the promise of more if they did not disrupt the long-standing agenda of the IARCs: to continue the industrialization of agriculture around the globe. If the IARCS expected Altieri and his colleagues to follow that script, however, they were in for a surprise.

In its opening statement, the NGOC critiqued the IARCs for expending tremendous resources on projects without consulting or involving local campesinos, overlooking their knowledge, experience, and priorities. Miguel and the NGOC members used their observations, however, to demonstrate a clear pattern: under conditions of increasing insecurity and ecological marginality, Indigenous production choices consistently outperformed the options promoted by the CGIAR.

In their report, Miguel and his NGO colleagues delivered a list of demands. The most critical points called on the CGIAR to: (1) ensure that the science is relevant to the needs of the rural poor; (2) recognize that NGOs are not willing to serve as extensionists of CGIAR technology and that there must be a collaborative process through which NGOs can participate in technology development from the beginning; (3) ensure that the CGIAR draws on successful experiences as determined by local farmers, focusing on yield stability in marginal areas as a basis for informing macro-level policies; (4) integrate natural resource accounting techniques to evaluate the environmental impacts of CGIAR field projects, including estimates of damages/ externalities associated with various technologies to guide future research; and (5) reorganize and reprioritize the CGIAR's budgeting for biotechnology.

This final point raised a fundamental challenge regarding the major emphasis that the IARCs placed on biotechnology research at the expense of an agroecological approach (e.g., natural resource management research projects) relevant to the needs of the rural poor (CGIAR 1997).

In the absence of an explicit response to the NGOC's recommendations, members of the NGOC later cited the CGIAR's continued emphasis on genetic engineering technologies as evidence of its apparent rejection of agroecological research and the role of campesinos and their communities in guiding IARC priorities (GRAIN 2003). By the early 2000s, Altieri, along with other members of the NGOC, had resigned from their positions. The CGIAR announced that it was reassessing the relationships between NGOs and its international centres.

Even as Altieri endeavoured to advance the role of agroecology at an international level, his ability to do so within the United States was becoming increasingly tenuous.

Note

1   One of the key people at Chapingo who arranged for this presentation at the CIMMYT was Tomas Martinez Saldana, a well-known faculty member of UACH and researcher focusing on the socioeconomic impacts of the Green Revolution in Mexico.

13 Agroecology at Berkeley: A Path Not Taken

Casual observers might easily interpret the demise of the Division of Biological Control at Berkeley as simply the march of progress, the common fate of disciplines that have failed to bring sufficient outside funding to sustain their research and professional standing. To many, its end seemed like nothing more than the triumph of new scientific approaches over outdated ones. The university had undergone extensive physical infrastructure changes. The off-campus, one-storied buildings where Altieri and his colleagues analyzed plant, animal, and human ecologies reflected an antiquated approach lacking the modern laboratories and computational power that increasingly characterized the university's capital-intensive community of researchers.

Yet a closer examination regarding this community of scientists reveals a far more complex and controversial history. In contrast to the simplistic narrative of old versus new science, the details of the Division's disappearance reflect a long and protracted conflict regarding the production of knowledge and its beneficiaries. A conflict embedded in the politics of science.

As the Division's longtime leader, Donald Dahlsten was constantly reminded that the work he and his colleagues pursued often put them at odds with powerful forces in California agriculture. A basic premise of their work, opposing the widespread application of pesticides, meant that many of California's major agribusinesses, as well as global chemical companies, found the Division's faculty to be a problem. Inside the university, faculty whose work was funded by these same industries also found the Division's research problematic — its ecological focus clashed with the priorities of an academic environment increasingly shaped by the agri-chemical complex that had long dominated the state's agriculture.

These conflicts also appeared at a more nuanced level with respect to foundational knowledge claims. I vividly recall that during one of the many occasions when Professor Dahlsten shared his office space with me, he remarked on the troubling shift he was witnessing. The careful field observations he had once made to understand ecological relationships were being displaced by abstract theoretical models. Equally troubling for Dahlsten was the trend among other researchers in the university to dismiss the negative consequences of industrial agriculture, including the poisoning of workers and their communities, the contaminants in food, water and soils, or the notion that an alternative scientific approach based on ecological and social imperatives should be embraced by a public university.[1] Dahlsten's reflections on the changes at Berkeley signalled that the influence of financial interests in defining the cutting edge of scientific projects had fostered a profound disconnect between increasingly reductionist approaches and independent assessment of their impact on nature, society, or the production of knowledge.

One particular example of this disconnect had emerged in a class I had instructed in the early 2000s with Professor Dahlsten. In class, Don had given his wholehearted support to a large group of graduate students who were challenging a proposed "gift" of substantial research funding by a major chemical corporation to the department. While the corporation maintained that the funding would be available to all faculty, the students argued that the gift would, in practice, elevate projects having commercial value relevant to the corporation's interests. Such a gift would inevitably alter research project topics and questions, the focus of courses, the recruitment and composition of faculty, and the nature of what students and faculty would come to understand as the purpose of their work. The graduate students were standing at the juncture to what Xolocotzi and others had warned about decades earlier. Behind the gift was the perennial question: "Whose science" would be pursued at Berkeley?

Dahlsten's support for the graduate students reflected a larger conversation that had permeated countless academic discussions over many decades. It was a conversation that had been sparked with the free-speech movement of the 1960s — a moment that had opened the doors on the long-term impact of private-sector interests defining certain principles regarding the production of knowledge. For Dahlsten, Altieri, and many of their colleagues the entry of corporate funding into the university was not a source for fostering revolutionary scientific breakthroughs, but a counterrevolutionary force to what truly independent faculty and researchers might pursue in a democratic setting.

Challenging New Projects in Science and Technology

True to their commitment regarding their integrity as independent researchers, in the early 2000s members of Berkeley's Division of Biological Control were drawn into yet another political fight surrounding the production of knowledge.[2] This time the fight involved an open critique of their colleagues across the university who were engaged in an increasingly valuable part of the university's portfolio: biotechnology. More specifically, the small band of faculty members who had for so long challenged the practices of major agribusinesses now took up a fight involving a novel agricultural technology: transgenic corn. Indeed, the new struggle centred on a technology being supported by many researchers working on various campuses of the University of California and the damaging consequences of their work for Mexicans generally, but especially Mexico's campesinos.[3]

The role of transgenic corn in many respects reflected a similar scientific trajectory as was introduced by the Rockefeller Foundation's earlier work in the 1940s and later efforts conducted by its Mexican successor, the CIMMYT. Paralleling the questions that had emerged regarding the earlier industrialization of Mexico's agriculture, studies at the beginning of the new century pointed to empirical evidence of biotech presenting detrimental social and environmental impacts across the Americas (Otero 2008).

Indeed, some of the strongest critiques of agricultural biotechnologies emerged in Mexico during the early 2000s:

> Transgenes were found in local Mexican corn varieties in 2001, setting off highly charged debates about the extent to which GM corn poses a threat to native varieties in the crop's center of origin, domestication, and biodiversity. At the time the cultivation and scientific testing of GM corn were prohibited in Mexico, yet corn imports from the United States, where there is no required labeling or separation of transgenic corn, included genetically engineered varieties ... the GM corn controversy raises questions about the fate of the peasantry in an era of corporate agriculture and globalization. (Fitting 2011: 3)

Fitting, like many other authors, analyzed the problems facing Mexico's people through the continued spread of industrial agriculture, both its older forms and its latest innovation — transgenic seeds. Similar to the pattern established by the Rockefeller Foundation, transgenic seeds reshaped not simply the organization of Mexico's agricultural production, but its social and political landscapes as well. The impact of trade regimes such as NAFTA, predicated on advancing capital-intensive agriculture, meant that Mexico's agriculture would continue to strip campesino communities of their ability to control their access to land and food while transforming them into "rural subjects" as either agricultural entrepreneurs who produce for export or an inexpensive labour force (Otero 2008: 4).

Unlike the earlier program of industrialization, transgenic seeds provided a more efficient technology for emptying campesinos from the countryside whether they participated in the changing economy or not:

> Given the technology's effect of increasing the productivity rate of growth and its large-scale bias, small peasant farmers are increasingly rendered inefficient and expelled from agriculture. The withering away of small peasants as cultivators has consequences not only for the social structures, but also for the endangerment of biological diversity, to the extent that peasants have been the "curators," so to speak, of [much] of the region's crop biodiversity. Therefore, while Latin American agriculture may be producing a greater share of exports of grains, fruits, and vegetables, their ability to feed their own people has decreased, thus becoming dependent on increasing food imports, or increasing numbers of people simply going hungry. (Otero 2008: 19–20)

The observation that maize cultivation represented "a dynamic process in which native maize varieties (criollos) are maintained and developed through exchanges between cultivators and between cornfields" echoed Xolocotzi and others from decades earlier as they called attention to the ecological knowledge possessed by Mexico's campesinos. As critics of transgenic maize argued, the displacement of millions of campesinos would result in the loss of genetic variability for maize and the vulnerability of catastrophic losses through their replacement with uniform commercial varieties.

The devaluation and loss of ecological knowledge and practices possessed by the campesinato signalled mounting pressure to empty the countryside of the very communities that had long provided resilience and the ability to restore otherwise vulnerable ecologies. The counterrevolutionary impact of transgenic maize promised to take away the lands that had been granted to millions of campesinos via a seemingly neutral technology fashioned by impartial scientists. A well-known Mexican anthropologist, Armando Bartra, referred to the right not to migrate, "the right to stay home" (Otero 2008: 5). Yet in the face of this new technology, the notion that Mexico's campesinos still possessed a right to stay home or to sovereignty over their food production was rapidly disappearing.

The conflicts surrounding the genetic engineering of maize in the early 2000s fully enveloped a large group of students, along with faculty and researchers with roots in the Division of Biological Control.

In 2001 Ignacio Chapella and David Quist from the University of California, Berkeley reported that they had found the presence of three distinct transgenic DNA sequences among local maize in the Sierra Norte of Oaxaca: the Bacillus thuringiensis (Bt) toxin gene, the cauliflower mosaic virus (CaMV) gene promoter, and the nopaline synthase (nos) terminator sequence.... When the study was eventually published in *Nature* it sparked an international debate about not only the risks of gene flow between such genetically modified corn and criollos but the reliability of the researchers' study. Although *Nature* had peer-reviewed and accepted the paper, in an unprecedented turn the journal withdrew its editorial support for the study in 2002.... The industry and government perspective that GM corn can be beneficial to Mexican native varieties also tends to view improved varieties, including GM seed, as a remedy to the inefficiency of "traditional" maize production.... As Fitting noted in 2011 and has been more substantially demonstrated in recent years, "high-yielding varieties are more "efficient" when the other costs — monetary, environmental, and social — are ignored. (Otero 2008: 50–4)

Quist and Chapela's findings quickly devolved into a reductionist fight regarding the methodology in tracking GM corn contaminants affecting native varieties. But as many others noted, the battle over what Quist and Chapela had discovered was largely not in dispute and has been amply demonstrated in subsequent years. The most important element for many Mexicans was the demonstration that campesinos were being threatened by the latest version of a technology predicated primarily on the extraction of profits from rural communities across the country.

For Quist and Chapela, the conflict wholly engaged the biotechnology industry because their investigation lifted the veil, making plain the politics of this purportedly neutral technology as the carrier to advance the latest version of an older and failing Green Revolution. Although Quist and Chapela acknowledged certain methodological errors in their work, their finding of transgenic DNA contamination of criollos was never refuted by later scientific studies. What was discovered was a smear campaign traced to a consulting firm with ties to a major international agribusiness. For many of their graduate students and faculty colleagues there was a lasting impression of being cautious about tangling with biotechnology firms. Ignacio Chapela experienced increasing marginalization in both his employment at Berkeley and career more generally.

Altieri analyzed the problem by situating it within a deeper and older issue: the industrialization of agricultural science.

Genetic engineering is another technological fix, a supposed "magic bullet," aimed at circumventing the environmental problems of agriculture, which themselves stem from an earlier round of technological fixes. Rather than questioning the flawed assumptions that created these problems, biotechnology advances single-gene solutions to challenges arising from ecologically unstable monocultures built on industrial models of efficiency. This unilateral and reductionist approach has already proven ecologically unsound in the case of pesticides, whose promoters once advocated a "one chemical–one pest" strategy — now echoed in biotechnology's "one gene–one pest" approach (Altieri 2004: xvi).

Beyond Berkeley, however, the GMO maize episode produced a much wider social response. Across Mexico, the threat to a crop that has long served as a national cultural icon mobilized unprecedented opposition in an array of social and academic communities and realms. While the agribusiness giant temporarily silenced some of the faculty and perhaps none of the students at Berkeley, the reaction in Mexico galvanized a nationwide rejection of industrial agriculture and transgenic biotechnology as well as opposition to hegemonic global trade practices and disapproval of health risks and genetically modified crops (Toledo and Barrera-Bassols 2017).

For Don Dahlsten, the legacy that he and his colleagues exercised to promote a science anchored in ecological and societal needs meant that their university positions were also becoming increasingly tenuous. Their perspective, coupled with a willingness to engage in projects and campaigns designed to not only support social justice but to openly criticize new technologies developed elsewhere in the university, placed them on a precarious path. The Division's full complement of ten faculty members — who once published prodigious articles, attracted substantial research funds, and gained national attention in the 1970s and 1980s — was, by the early 1990s, reorganized and its faculty subsumed into a larger research unit. By 1995 the remaining four faculty members from the original group were no longer a sufficient number to reproduce their previous team as a cohesive unit. The retirement of Professor Altieri in 2023 effectively marked the demise of the team that had worked so hard to promote an alternative to California's chemical-intensive agriculture (Jennings 1997).

The story of agroecology, however, does not end there. The faculty members with the Division of Biological Control, in fact, were very aware of the considerable time and effort that Altieri had expended on growing agroecology in other places across the Americas. Indeed, one of Altieri's closest colleagues, Professor Emeritus Peter Rosset, has mentored generations of agroecologists across the Americas from his position as a faculty member at ECOSUR in Chiapas, Mexico. Rosset's works, like a number of other agroecologists, could easily comprise a parallel publication — which probably explains why Fernwood Publishing has published a major work co-authored by Rosset. While Rosset and Altieri have collaborated on numerous projects over the years, it may be fair to characterize the former's endeavours as more focused on linking the science of agroecology to social movements, with a special emphasis on LVC.

The voices now carrying agroecology forward are numerous and include not simply academics, but environmental and social justice activists and advocates, Indigenous communities, members of farming cooperatives, labour organizers, journalists, feminist organizations, consumer groups, small farms and businesses, and, of course, legions of campesinos who remain committed to agroecology while not even belonging to any organizations. One of the fascinating aspects of the expanding profile of agroecology is the vast number of people across the Americas and beyond whose work in the world corresponds to precisely what Rosset, Altieri, and so many others have sought to support and advance — recognizing one's revolutionary power in society.

The concluding chapter provides but a glimpse of the multifaceted array of agroecological works in select parts of the Americas. These passages are not only encouraging about the future of agroecology; the infrastructures they detail also demonstrate the compelling dimensions of a revolutionary science.

Altieri's former colleagues, those no longer with us, would not be surprised to see that the legacy of their efforts to build an alternative to industrial agriculture has continued to expand across the Americas. Indeed, the stature of agroecology and agroecologists has only increased over time.

Meetings of agroecologists in many parts of the world continue to draw hundreds, sometimes thousands of people eager to pursue the widening array of global initiatives. This sustained, large-scale public interest in fashioning a new agriculture, with the assistance of a community of scientists, serves as a counterpoint to the tragedy that unfolded at Berkeley. While the conclusion of the conflicts surrounding the struggle to establish agroecology has yet to be written, this nascent science and its linkage to a political movement suggests the resilience of an alternative scientific mission; one that begins with projects organized and directed by those at the bottom.

Notes

1 This and other references to Donald Dahlsten are derived from the numerous conversations the author had as a visiting scholar with the Department of Environmental Science, Policy, and Management at UC Berkeley from 1997 to 1999.
2 The Division of Biological Control was reorganized as part of the larger Department of Environmental Science, Policy, and Management, a move that would diminish its ability to operate as an independent group of researchers as well as impacting their control over the replacement and recruitment of faculty and access to resources.
3 The reader should note that many of the passages of this story are taken from the excellent writings of Elizabeth Fitting (2011).

Part IV

Conclusions: Linking a Science and a Movement

While agroecology has origins in traditions and practices initiated many centuries ago, a pivotal moment sparking its emergence as a science occurred with an organized response to the Rockefeller Foundation's dissemination of industrial agriculture throughout the Americas. Today, even as industrial agriculture continues to dominate many landscapes, agroecologists are increasingly recognized for advancing a compelling alternative. In the twenty-first century, they have established a new science built on the systematic integration of campesino knowledge and skills, alongside the expansion of peasant participation and political rights. The concluding chapters, focusing on infrastructures of support and infrastructures of resistance, summarize some of the approaches for advancing both the science and the movement.

Yet, as agroecology expands in Mexico, Central America, and South America, difficult questions remain. What does agroecology mean for those living far from campesino communities? Beyond celebrating its successes, the closing reflections raise broader challenges: What lessons can be drawn from the struggles of agroecologists and millions of campesinos? How should we interpret these lessons in light of the history of the Green Revolution and the politicized science of industrial agriculture? What are the implications for leaving industrial agriculture largely unchanged while trying to create the conditions necessary for the dramatic expansion of agroecology? And perhaps most crucially: what does it mean to leave unchallenged the political interests and structures that sustain industrial agriculture?

14 Agroecology and Infrastructures of Support

Over a few short decades, Latin America's agroecologists have made impressive strides — they have forged deep and enduring alliances between scientists, campesinos, and countless others to recognize and celebrate the traditions, knowledge, and daily work of the more than 65 million campesinos, of whom some 40–55 million are members of various Indigenous groups speaking over a thousand languages. Integral to providing food sovereignty for millions are an array of partnerships and affinities, including countless participants engaged in maize fairs, family gardens, designing and demonstrating polycultures, maintaining seed banks/exchanges, conducting research and demonstrations of new practices to enhance biodiversity, promoting village schools, sponsoring campaigns, laws and policies, building new relationships with urban partners, engaging in environmental restoration projects, connecting producers and consumers, developing alternative markets, and encouraging a larger solidarity movement based on a commitment to agroecological principles (Toledo and Barrera-Bassols 2017).

Agroecology's focus on traditional campesino practices offers a unique scientific approach — one that prioritizes the knowledge of agricultural workers as foundational for understanding what truly works in particular places and different landscapes. It establishes a methodological difference from industrial agriculture, but also a political one — with a framework where workers are integral partners in the production of knowledge, embedding a democratic process into the very fabric of defining problems, developing technologies, and shaping the objectives of scientific missions.

By expanding the scope of agricultural inquiry, agroecology offers a more comprehensive understanding of environmental and social processes and impacts. It is an orientation that avoids the silo effect of only examining isolated metrics, such as crude yields per land unit, while ignoring the long-term ecological as well as social impacts of agricultural interventions. In this way, agroecology avoids the historical and systematic pitfalls of industrial agriculture, in which everything beyond increased production becomes an "unintended" or "unforeseen" consequence.

This orientation of agroecology enables engagement with a broader array of projects as well as partners; the latter including an array of interdisciplinary scientists along with allied communities. These other projects and partnerships demonstrate the depth and reach of agroecology for addressing interconnected global challenges, revealing its potential to extend beyond the boundaries of agricultural fields and increasing understanding of the value of a larger infrastructure to support related projects. These infrastructures, while sometimes physical in nature, also reflect a distinct set of guiding principles, rooted in the social movements that support and define agroecology.

Agroecology's infrastructure distinctively emerges through its interdisciplinary integration of knowledge and research findings, as well as its critical assessment of the legacies of industrial agriculture and the politics of science. This critical perspective has inspired innovative research inquiries, particularly exploring not only the ecological dimensions, but the pivotal role of labour within agricultural systems. Agroecological research comprehensively examines production systems by interconnecting multiple dimensions: agricultural workers' experiences, ecological impacts, public health outcomes, nutritional implications, land rights, and market access. It operates across temporal and spatial scales, the former especially by honouring historical knowledge and practices.

This orientation explains why the practical applications of agroecology so often look different from those of industrial agricultural science. Indeed, agroecological research looks very different from that of either projects sponsored by the Rockefeller Foundation years ago or its modern counterparts, supported by such groups as the Gates Foundation, the International Maize and Wheat Improvement Center, or the Consultative Group on International Agricultural Research. These entities largely focus on a narrow set of agricultural production techniques divorced from their larger social, political, and ecological consequences.

Contemporary agroecological projects are now energizing a generation of students, activists, and rural workers across the Americas. It is an exciting moment to observe the enthusiasm of so many working to construct an alternative to industrial agriculture's legacy of depleted aquifers and poisoned waterways, massive soil loss, highly processed foods, species extinction, and production practices that worsen climates and ecologies on a global scale.

Agroecologists often speak of designing an agriculture to improve a range of ecosystem services, such as classical biological controls and the use of one insect species (e.g., a particular kind of wasp) to control the population of another insect (e.g., caterpillar). Almost by definition, agroecological practitioners collaborate with campesinos to understand and cooperatively design adaptive landscapes capable of withstanding increasingly volatile conditions, both climatic and economic. Beyond the ecological level, these collaborations extend to discussions about achieving broader social objectives, such as how to provide food, shelter, and expanded wages and opportunities for hundreds of millions of humans, who can preserve and enhance landscapes on a vast scale as a service to humanity. These dynamics also foster food sovereignty, enhance cooperative and regional economies, and more fully recognize the value of the skills and knowledge possessed by campesinos and the vast number of other marginal and unemployed workers in rural and other settings.

It is in both the learning from campesinos about traditional ecological practices as well as supporting them in pursuing larger social objectives where it becomes evident that Latin America's agroecologists are not engaged in some variant of industrial agriculture. Their work reflects an entirely different science, a revolutionary science, on that has become integrally linked to the struggles large and small, with an aim to advancing the standing of hundreds of millions of campesinos across the Americas.

The examples presented below illustrate some of the agroecological infrastructures appearing across the Americas. As mentioned, these examples illustrate only a small part of the economic, social, cultural, and political facets of agroecology as a science, practice, and movement.

Agroecological Lighthouses

Among the many people celebrating agroecology is scientist Augustin Infante, the longtime director of a recognized centre demonstrating agroecology's potential as a transformative alternative to industrial agriculture. Established in the early 1990s, the demonstration site near Yumbel, a town located outside of Concepción, Chile, serves as an enduring example of how many agroecologists demonstrate their work to a broader public. Referred to as "faros" or lighthouses, the site embodies a strategic campaign to connect campesinos' traditional agricultural knowledge with urban populations increasingly interested in ecological and traditional farming approaches.

The lighthouse at Yumbel, in certain respects, illustrates the differences between circular versus linear models of farming. Whereas linear techniques resemble an industrial model of monocultures heavily reliant on capital-intensive inputs (e.g., synthetic fertilizers, pesticides, irrigation) and operating without regard to impacts on soils, society, or environmental health with the sole criteria of increasing production, circular farming is designed to attain production in the context of multiple considerations, including working within the confines of available resources (e.g., ecological services, saved seeds, nutrient and carbon recycling) to sustain and nurture the livelihoods and well-being of campesinos and the surrounding territories.

Prior to the COVID-19 pandemic, Augustin and his team hosted weekly meetings to guide (and feed!) dozens of faro visitors through their many hectares of fruits, vegetables, and herbs. Participants represented a diverse cross-section of rural life: campesinos, retirees, and working people, many with small landholdings or access to increasingly rare communal properties across Chile.

The lunchtime conversations echoed themes expressed by others holding such discussions across Latin America: personal health, access to healthy foods, and a different way of farming than what was generally practised by the larger agribusinesses up and down the major valleys. Throughout these conversations was an ever-present theme, widely accepted among those showing up at Yumbel: behind the facade of Chile's modern agriculture is growing evidence of its deep flaws and worsening failures.

Augustin's farm tour began by showing photographs of the original wheat fields, cultivated since the early twentieth century. By the 1940s, decades of farming a single grain across these hillsides had resulted in lands bereft of their original richness — bare soils, gullies, and worn-out acres unable to yield anything, including the previously declining harvests of wheat. The contrasting landscape, now present for all to see, was stunning. The lands reflected renewed health, backed by a running record of statistics measuring everything from soil fertility and moisture levels to the growing diversity of plants and other indicators of ecological well-being. On another level, one of the most vibrant transformations visible in the fields at Yumbel is in those gathered, as they express excitement and pride in pursuing a collective dream: engaging in a common endeavour that promises to leave the Earth and our communities better than they found them. And all in the service of providing their fellow humans with nutritious, healthy food.

Agroecology and Community Restoration of Watersheds

Rolando Rojas is a quiet, humble Chilean who has worked for many years in southern Chile. By his resume, one might mistake him as simply another undistinguised member of that country's class of professionals. Yet, like so many younger Chileans, Rolando refuses to be another cog in a neoliberal economy designed decades ago by a group of economists from Chicago.

Over lunch on Chiloé, his reserved demeanour vanishes as he animatedly describes his involvement in an agroecological project — a community-driven initiative that might easily go unnoticed but holds immense significance.

Chiloé is a magical place. As one in a chain of islands in southern Chile, Chiloé has strong traditions of protecting native species, saving and exchanging seeds, and merging agriculture with the surrounding ecology. The celebration of these facets of agroecology at one of the island's central plazas reflects popular support. Having originally visited Chiloé more than thirty years ago, it is also evident that its magical qualities are vanishing, reflecting the immense reach of industrial agriculture.

In recent times, the same private-sector policies for advancing industrial agriculture have extended their reach to encompass Chile's forests. The consequences are immediately recognizable in the stands of eucalyptus trees now populating the island. These industrialized monocrops have not just displaced the lush diversity of Chiloé's flora; they have also ushered in a new crisis in local watersheds and overdrawn water tables. As a result, residents now face a threat largely unknown previously: a lack of water.

It was in this context that I met Rolando, a member of a new generation of agroecologists, a group I refer to as agroecology's "fix-it team." One morning in Ancud, Rolando shared details of a recently completed project in one of Chiloé's many small communities. The problem confronting this isolated community was, as mentioned, their recent loss of water — something their members learned could not be answered by simply drilling more or deeper wells. Indeed, their initial conclusion was that in the absence of an expensive public project to deliver water from a great distance, they would have to simply abandon their homes and move elsewhere.

When Rolando learned about the threat to this community, he and a team of fellow agroecologists quickly surveyed the problem, and they were able to provide the community with an arguably much better alternative. Devising a plan for reconfiguring their landscape with a different set of plants and eliminating the industrial trees, they then focused on capturing and retaining various water sources to provide permanent water resources.

Roland's explanation, however, highlighted a specific characteristic of agroecological projects. The ongoing success of the newly designed landscapes depended on the community learning how to manage and sustain these practices. Unlike a typical government-sponsored infrastructure project, Rolando and his team called on community members to become active participants in the planning, building, and maintenance of the project. The reasoning was clear: vesting the community with a knowledge about agroecological designs would provide them the foundations for understanding how to monitor and adjust community practices necessary to maintain this invaluable resource into the future.

A second, more subtle quality emerged, too — the importance of community-based control over this essential resource. In other words, it was crucial to foster a democratic space where local people could exercise power over their economy. The expansion of industrial forestry and its impact on water resources across Chiloé presents an existential crisis for many of its residents. Advertisements for "well-digging services" are just one of many signs of the growing contradiction between free-market policies and community survival.

It is within this complex landscape that Rolando and his colleagues navigate an especially difficult task, as part of a movement spreading across the continent: demonstrating the value of applying agroecological principles to solve one of the most pressing issues for countless communities — access to water. Another, perhaps even more difficult task, involves encouraging individuals to act together to achieve a common purpose. Particularly in Chile, given the legacy of imposed free-market regimes (e.g., private control of property and natural resources such as water), a cohesive and unified community is not a given. In this way, the political context surrounding democratic institutions and infrastructure supporting community-controlled decisions (e.g., properties and rights held in common, such as ejidos) is extremely important.

Agroecology and Pedagogical Infrastructure

A recent gathering sponsored by the National University of Colombia in Palmira reflected what has become a growing trend among agroecologists in Latin America: virtual meetings accessible to larger audiences across vast geographies. This event also reflected another trend: nearly all of the diverse speakers were women, many of them representing a younger generation of agroecologists.

After several decades, one might assume that the legacy of the Green Revolution had long since passed, replaced by new forms, new technologies, new understandings that had moved well beyond its original mission in Mexican agriculture. Yet watching Bolivian doctoral student Adriana del Rosario Rodriguez R.'s YouTube presentation concluding a year-long course offered at a well-known Colombian university in 2022, one discovers that the Green Revolution is explicitly referenced as an integral part of the instructional material — though primarily to illustrate the pitfalls of research methods rooted in the practices of industrial agricultural science.

In the 2022 course, Adriana, with the Universidad Católica Boliviana, gave a video presentation on designing agroecology curricula that featured detailed discussion on research methods and pedagogy. In defining the essential elements of her pedagogy, Adriana underscored the vital importance of "territorialization," which for Adriana locates an important body of knowledge in the lived experiences and knowings of local campesinxs (terminology she uses to refer to people of all genders). This knowledge is based not only on an understanding of the physical landscape of plants, biological interactions with insects and animals, yields of various sorts, and so on, but also on an understanding of obstacles to their livelihoods, be these climatic, political, or economic.

A striking aspect of Adriana's presentation on a pedagogy for agroecology and its basis in campesinx knowledge in a given region appeared in her very first slide, where she carefully explained how the Green Revolution model of "extensionismo" actively undermines the larger objectives of agroecology. In the twenty-first century, it is apparent that for many agroecologists working on the ground in the Americas, the Green Revolution's legacy continues to threaten the emergence of a new science (Rodriguez 2022).

The lineup of speakers for the course offered by la Universidad Nacional de Colombia in Palmira stood out in several ways. Going back to the early days of the Rockefeller Foundation but even extending into contemporary times, it feels incredibly rare for a Bolivian woman completing her doctorate to be a featured speaker at a conference where most presenters and organization directors were also women. This contrasts starkly to the male-dominated scientific meetings that persist worldwide. Yet even this remarkable diversity was exceeded by the purpose of this gathering, made explicit by its title: *Agroecology as a transformative science for achieving food sovereignty.*

It is important to understand that Adriana's statements link to another, larger facet of agroecology and the merging of its scientific work with the practice of building food sovereignty. As Peter Rosset has explained, it is specifically because of the horizontal sharing of knowledge — between, for example, campesinos and scientists — that food sovereignty represents both a strategy and an on-the-ground practice for bringing together both those who work the fields and those who elucidate the science of agroecology. This knowledge sharing among equals (often referred to as diálogo de saberes) represents an essential part of education — not simply the top-down perspective of university experts, but from the bottom up, with the creation of agroecology schools based in the communities of campesinos. The larger regional and international meetings of agroecologists, such as those organized by LVC, also provide fora for discussing and deliberating on such matters of food sovereignty as discussions of equals.[1]

As the reader might recognize, these educational structures used in many parts of the Americas are reminiscent of the early peasant schools initiated after the Mexican Revolution but obliterated by the Rockefeller Foundation's emphasis on top-down transmission of knowledge (and authority) according to a model of industrial agriculture, totally divorced from the knowledge and desires of campesinos. In a dramatic reversal of that history, peasant schools of agroecology have re-emerged in many parts of the Americas and beyond. The blossoming of campesino-based schools now extends from villages to universities in many parts of Latin America. Indeed, some of the best known campesino-oriented programs at major universities are recognized as Latin American Institutes of Agroecology or IALAs. These schools set in motion a basis for campesinos to collectively exercise political power over their lands and lives and to unite with other communities as a social movement. In contrast to the prior Green Revolution model of experts instructing trainees solely focused on methods of increased production, the IALAs and campesino schools approach agroecology as a fundamental tool for constructing a new relationship between people and nature. In the words of Peter Rosset and colleagues, "The graduates of peasant schools are people capable of transforming power relations and promoting structural changes that allow their societies to approach the realization of food and popular sovereignty" (Rosset et al. 2019).

Agroecological Infrastructures of Food Sovereignty

One of the glaring features of industrial agriculture across the Americas is the large number of people who go hungry. The numbers represent a substantial population of not just homeless people, but a much larger group of impoverished, jobless, or marginally employed as well as elderly with insufficient incomes. While state and federal programs combined with food banks have distributed excess food to provide some temporary support, such programs have become increasingly restrictive and less accessible. In the United States in 2025 it appears likely that federal laws and national programs ensuring modest measures of support for adequate nutrition will substantially diminish.

Even for working families with greater incomes, access to fresh fruits, vegetables, and nutritious foods are frequently beyond their means. Compounding this situation is the fact that many food crops, especially ones grown as luxury foods for wealthy households (e.g., berries, avocados, cherries, wine grapes) frequently use scarce resources (water, capital, energy), further marginalizing the affordability of basic foods for most consumers. While this situation is often regarded as a "Third World" problem, increasing numbers of consumers across the Americas confront these same barriers to food access, availability, and affordability.

Many households as well as social organizations have become increasingly supportive of food sovereignty. It is no surprise that agroecologists have been drawn to and participate extensively in the development of food sovereignty, recognizing that it reflects many of the basic purposes of agroecology, As Nicholls and Altieri explain:

> Increasing the productivity of urban agriculture contributes to local food security by enhancing the ability of households to access food and improves nutrition by increasing the ability of families to diversify diets. It is likely that urban food production will expand as more people realize that in times of crisis, access to locally produced food is strategic ... can fortify people's immune systems ... and help the intestinal "good" bacteria and the overall gut microbiome health. (Altieri and Nicholls 2020)

Altieri and his colleagues note the importance of locating agroecological projects involving food sovereignty not simply in rural areas, but especially in urban areas: The same well-established agroecological principles used in rural areas for the design and management of diversified farms where external inputs are replaced by natural processes can be applied to urban farms.

Agroecology and Alternative Market Infrastructures

In certain respects, the expansion of agroecology mirrors the process once pursued by the Rockefeller Foundation decades ago, only in reverse. Rather than focusing on top-down, industrial-scale agribusiness, it begins with the people most directly involved in agriculture. It is a process designed to foster alternative market arrangements, extends political power to the currently powerless, builds physical infrastructure to support agroecology on a much larger scale, and broadens agricultural knowledge and practice to include the relationships between agriculture and ecology, food production and health, and a wide range of social dynamics.

A strategic place to initiate this work, identified by several agroecologists, begins with expanding the infrastructure for popular economies. Popular economies, and similar terms used by various Latin American authors, refer to informal and less recognized channels of commerce. Popular economies are extremely important for linking campesinos and their urban counterparts to larger regional markets.

In much of the Americas, this economic sphere reflects an economy where exchanges and services operate without access to so-called modern, global finance. The network of popular economies is characterized by infrastructures of transport and marketing of agroecological goods frequently having particular importance in the lives of many Americans. These linkages provided by popular economies serve what might be the most essential of all purposes for campesinos and their urban counterparts: food sovereignty.

One illustration reflecting the role of popular economies can be found on the outskirts of Santiago, Chile. In a series of projects extending over several years, different *barrios* around the city Santiago have held discussions regarding the needs of their communities. In one such *barrio*, following a series of public meetings, a consensus was reached: among their highest priorities was the need to access affordable fresh fruits and vegetables. Decades of modern commerce had left their community chronically marginalized to the operations of modern agribusinesses.

As residents became aware that a distant *barrio* possessed numerous small, agroecological producers, they crafted a proposal to establish a more permanent trade relationship — one in which fresh produce could be regularly supplied in exchange for services their community could offer in return. Two Chilean academics who assisted in forging this linkage underscored that the most important part of their work was simply to facilitate the discussions between the two barrios and to respond to questions about how to surmount some of the existing obstacles. Indeed, among the many benefits of this project was the realization that both communities possessed valuable resources among their own residents — lawyers, transport workers, accountants, and others — to fashion what is referred to as a horizontal relationship. For many agroecologists, these relationships have also been identified as campesino to campesino. This example presents a potentially more expanded opportunity of forging horizontal alliances between campesinos with other workers, especially what might be recognized as their urban counterparts (Peredo Parada 2024).

As exciting as this project is for the residents of these two barrios, there are still other agroecological efforts pursuing various pedagogical purposes (Gago et al. 2023). Some of these projects focus on linking campesinos with others who provide a kind of physical or service infrastructure to transport their products to market, to communicate and organize with other communities relating to trade, to create regional water, transport, and other infrastructure, or to instruct university researchers about the needs of their communities.

The questions arising in these instances are especially important for building alliances in other parts of the Americas, such as among Canadian agricultural workers and others. Canadian alliances come readily to mind precisely because of the political campaigns that have been advanced to link Canadian consumers, farmers, and farmworkers. Indeed, Canada's National Farmers Union (NFU) work is distinctive in terms of building alliances among agricultural workers, farmers, consumers, Indigenous people, and other communities in promoting alternatives to markets dominated by agribusinesses (Desmarais 2019). NFU efforts encompass solidarity campaigns fostering linkages with campesino-based organizations in other parts of the world, such as LVC.

Another element of the Canadian experience is the growing awareness of the continuing encroachment and threat of industrial agriculture on Indigenous peoples, particularly in light of a changing climate. For many of Canada's Indigenous people there is simultaneously a sensitivity about the precise meaning of "agroecology," as some understand this term as merely a set of techniques devoid of any social, political, or ecological context. In keeping with an emphasis so often expressed by communities elsewhere, certain of Canada's Indigenous peoples have suggested that agroecology should serve as a starting point for a conversation that extends to larger issues including food and land sovereignty (Jewell Price 2022).

In a very tangible sense, these approaches represent the opposite of the market structures that were fostered and expanded by the Rockefeller Foundation decades ago. From the perspective of agroecologists working in Chile, one of the most exciting opportunities begins with building projects among those communities and producers who were never part of the design of a Green Revolution. This alternative agroecological model challenges the landscape shaped by an entirely different vision of agriculture.

One of the most remarkable characteristics of agroecology is its practical orientation. Unlike too many critiques of contemporary capitalism as operating at a level of lofty abstractions, agroecology presents a set of principles synced with a variety of practices based on demonstrated and measurable achievements. In many respects these constitute a plan for advancing both an alternative to industrial agriculture and a template for a twenty-first century economy — one that integrates ecological and human health, production within the context of those working the soils, and a governance structure that extends democratic rights to those who are currently excluded.

The Limits of Supportive Infrastructures

Recent decades have seen a robust demonstration of agroecological works across the Americas. The impacts of agroecology can be found in village schools, university programs, market cooperatives, seed exchange networks, alternative markets, consumer networks, urban gardens, popular gatherings, and demonstrations. One of the most notable networks in Mexico, Red Mexicana de Tianguis, has brought together many groups to engage in organizing, fundraising, political action, and collaborative efforts to promote a range of agroecological projects. Despite the popularity of so many expressions of support in communities across the Americas, many agroecologists note that the attacks on agroecology at all levels are a constant challenge (Astier et al. 2017).

One of the most striking visual experiences of visiting agroecological projects over the years is witnessing them in adversarial surroundings. Augustin's demonstration site in Yumbel, for instance, is noteworthy not only for its constant flow of interested participants from the surrounding region but also for being encircled by industrial agriculture, industrial aquaculture, and an export-driven regional and national economy designed primarily to serve foreign consumers. For the dominant industrial and financial interests, their concern about their impacts on Chilean ecologies and environments, workers and communities, fisheries and fauna, regional and local economies, and any semblance of democratic process or economic alternatives could not be less. Which is not to dismiss the value of the agroecological demonstration site at Yumbel. At the same time, the impact of the agroecological lighthouse in a landscape of dark forces requires something more than a compelling narrative about the heroic struggle of agroecologists.

A similar contrast exists in the places where agroecologists have thus far found refuge: universities and academic research institutes. Once again, we can recall the progressive marginalization of Altieri and colleagues at various places across the Americas. By comparison, their colleagues in the science of industrial agriculture are clearly receiving robust funding, office suites, travel budgets, and a full complement of faculty, teaching assistants, and support staff. As the story of agroecology and its demise at UC Berkeley illustrates, agroecologists remain an endangered species in many settings.

The threats to agroecology as a science only reinforce its critical role in a larger social movement for overcoming the dominant role of industrial agriculture around the world.

We see agroecology as a key form of resistance to an economic system that puts profit before life.... The real solution to the crises of the climate, malnutrition, etc., will not come from conforming to the industrial model. We must transform it and build our own local food systems that create new rural-urban links, based on truly agroecological food production by peasants, artisanal fishers, pastoralists, indigenous peoples, urban farmers, etc. We cannot allow agroecology to be a tool of the industrial food production model: we see it as the essential alternative to that model, and as the means of transforming how we produce and consume food into something better for humanity and our Mother Earth. LVC (Rosset and Altieri 2000)

Note

1   See especially Peter Rosset, presentation at Food Sovereignty Conference (2015): <youtube.com/watch?v=UyLZ4I6EguI>.

15. Agroecology and Infrastructures of Resistance

The setting could be described like any other evening gathering around a campfire. Just beyond the fenceline of a nearby farm, a fine meal had been cooked over the open flames. People had enjoyed enough of a very decent beer to engage in lively and good-natured conversations. A couple of guitars had appeared and the group was relaxing, listening to the music and to one another. It was like a gathering anywhere, except it wasn't — it was an unusual time and place.[1]

This was Cuba in 1996. During a time known as the "Special Period," the recently expired Soviet Union had withdrawn its economic support and the island nation was on its own. It was a moment when life for Cuba's inhabitants had been turned upside down. Years of fully subsidized petroleum products, which had powered a mechanized agricultural system, had come to an abrupt halt — and with it, food production had become much more precarious. The conversation around the campfire, like so many happening across Cuba, turned to the increasingly urgent question of how to grow food.

Earlier that day, three young farmworkers had detailed how they carried out their work in the absence of petroleum inputs. As the tour of their fields concluded, they next provided a wonderful meal, later accompanied by the sharing of guitars and song around the campfire. Over the past several days, many of the Cuban agricultural workers had demonstrated an impressive understanding about farming without capital-intensive inputs, including how to improve soil fertility without synthetic fertilizers, providing popular foods while avoiding ultra-processing, and extending food production to urban gardens. Like so many others we had met on this visit, everyone was hungry to gain more knowledge about how to do a better job of feeding their families, friends, neighbours, and even strangers from foreign lands.

Among the group of strangers gathered around that night's campfire was an especially important group of guests — at least in the eyes of the Cuban agricultural workers. The group of roughly fifteen people included three farmers from California. Their presence bespoke their global reputation:

California's farmers were readily received as among the most successful in the world. It was in this context, of one farmer talking to another across a fire late in the evening, that the young Cuban asked his Californian counterpart the question that he had been waiting all evening to ask: "Knowing the hardship and difficulties we have faced in recent years, what would you have done?"

The question was neither antagonistic nor sarcastic. Rather, it was a genuine search for an answer he eagerly wanted to know. On one hand, the young Cuban was convinced that he and his companions' efforts were sound and following the right path; however, he was also keenly aware that perhaps these successful farmers from California had a superior approach they could share. Here was a moment, away from any official meeting, where these farmers could speak honestly with one another.

The California farmers hesitated, exchanging glances. Then, the senior member of their group did what is so common in conversations spanning different parts of the world — he repeated the question. "So you're asking if we were confronted with the sudden loss of petroleum, how would we respond?"

The Cubans nodded. It was clear from the looks on their faces that what came next was an answer they were wholly unprepared for.

"Oh, that's easy," stated the older California farmer, "... we'd starve!"

Years later, the impression that there is no practical alternatives to industrial agriculture is still perpetuated with myths and misunderstandings, despite being disproven by vast numbers of people who have maintained livelihoods for millennia without being immersed in a highly capitalized and petroleum-laden model of industrial agriculture. As one agroecologist has noted, campesinos, small farmers, community-based gardeners, and others use only 30 percent of the world's resources to provide 70 percent of the planet's food for billions of people (Shiva 2016a: 1). This led her to conclude that it is agroecology, not industrial agriculture, that feeds the world.

At this point in the twenty-first century, the question is no longer whether agroecology holds value as an alternative; rather; it is about helping a larger public comprehend the threat industrial agriculture poses to human civilization and the planet.

Misrepresenting the negative attributes of industrial agriculture lessens a public demand for alternatives. In the modern era this is referred to as disinformation or more insidiously as the destruction of knowledge. It is a phenomena recognized several decades ago by Michael Dove when observing those actively ignoring the value of a historical farming system (swidden) as "the political economy of ignorance" (Dove 1983: 85–99). It is particularly this divide between agroecologists and industrial agricultural scientists that indicates strikingly distinct trajectories for society and nature in the twenty-first century.

We are now in a moment where the links between industrial agriculture, its reliance on petroleum, and its cataclysmic consequences can only be sustained by actively distorting what is unfolding before us. As foretold by those gathered around a campfire in Cuba, we have entered a period of crisis; one integrally linked to the prevailing model of industrial agriculture.

The Political Economy of Ignorance: Industrial Agriculture and Its Consequences

A central feature of industrial agriculture in the twentieth century was its reliance on petroleum-based fuels and agrichemicals. The vast array of agrichemicals has included many petroleum-based pesticides, fertilizers, and plastics whose ingredients and waste byproducts have been found in human blood and organs, including the heart, brain, and testicles (Nehart et al. 2025). Even as many of these materials contain documented hazards, only negligible actions have been taken to address the damaging neurological, reproductive, and cumulative health effects on workers, their communities, and consumers arising from multiple and chronic exposures, as well as impacts on flora, fauna, and larger ecologies (US Centers for Disease Control and Prevention 2024).

Beyond the general exposures to agrichemicals in workplaces and the environment, dietary exposures to many of these same chemicals is also a concern. Adding to the realm of pesticides and other agrichemicals appearing as contaminants in food and water, there are also many classified as toxic air contaminants. While the sources of plastics in the body are still being investigated, many appear to originate with plastics in both raw and processed foods.

Even more urgency has been advanced recently to dramatically reduce industrial agriculture's reliance on fossil fuels and agrichemicals in order to reduce climate damage (Fossil Fertilizers Report 2022). Industrial agriculture accounts for approximately 20–30 percent of global carbon emissions. Scientists have long called for drastic reductions in its carbon footprint, advocating for a shift away from fossil fuels in planting, harvesting, and food processing, as well as the shortening of global supply chains while crafting positive policies such as supporting local food production to serve regional markets.

Overwhelming evidence underscores the urgency for adopting dramatically more restrictive approaches on the extraction, processing, use, and sale of petroleum-based chemicals and byproducts (Tickner 2003), and many governments have been (slowly!) implementing legislation to this end.

However, in places like the United States, we are now seeing dramatic reversals, with the dismantling of laws, regulatory agencies, and international treaties designed over many decades to transition away from petroleum-based production. The US in 2025 is at the forefront of this counterrevolution, led by a new oligarchy representing major economic sectors. Global technology companies in particular are viewed as increasingly complicit with spreading disinformation regarding the severity and gravity of fossil-fuelled crises. The dismantling of public institutions now extends to many government agencies and interdisciplinary university-based research centres that had documented the necessity of altering production practices reliant on petroleum and other fossil fuels. Is this evisceration of knowledge regarding the imminent threats facing humankind a temporary moment of irrationality? Or is this the evolution of capitalism toward a political economy of ignorance?

Accumulating evidence strongly indicates that the fossil-fuelled crises will become an overarching threat to the survival of human civilization by the end of this century (IPCC 2023). It is likewise becoming increasingly clear that the concentrated wealth and power of petroleum firms and agribusinesses are antagonistic to many democratic forms of governance. Campesinos and others seeking rights over land and water, food sovereignty, and to create alternative economies supporting their communities run counter to the design of industrial agriculture.

These cascading crises further erode the already dubious benefits of industrial agriculture while urgently prompting an alternative. It is at this juncture where the expanding achievements of agroecologists pose a distinct threat to the viability of industrial agriculture. The overwhelming findings gathered by scientists studying climate, oceans, plant and animal habitats, biodiversity, and so on only affirm the value of agroecology. Agribusinesses, chemical companies, and their allies have responded to this tide of damning evidence through one of their few alternatives: to support political regimes that will purge whole communities of scientists, journalists, and others who are documenting the host of negative consequences that result from petroleum and its heavy usage, including in industrial agriculture.

Joining and Creating Movements of Political Resistance

It might seem only natural that many peoples across the Americas should share a common movement to end oil — except for a not-so-small matter of democratic practice.

Is it appropriate to proscribe political actions on behalf of campesinos and others without their prior dialogue and agreement? This is a vitally important question, especially at a time of co-optation of agroecology by international agricultural research centres, agribusinesses,[2] and others are actively engaged in such schemes. In a variety of ways, groups such as LVC have already engaged in dialogues and statements signalling their advocacy that industrial agriculture poses a problem for people everywhere. Even so, it is important to not prescribe actions and priorities for others, most especially the hundreds of millions of campesinos who have already been subjected to decisions made for them by others.

Devising the political means to achieve agroecological practices, as articulated by one of the largest organizations representing campesinos across the world, LVC, is an integral step in linking the science of agroecology with the social movement supporting it. Agroecologists recognize that a social movement is essential to marshalling sufficient political power to move beyond petroleum, to achieve land and food sovereignty, and to transition to a postindustrial agriculture (Val et al. 2019). Even with an appreciation of agroecology's linkage to social movements, however, many questions remain.

The mass of humanity must confront a challenging reality: how to overcome the entrenched power of an industrial agriculture that has become fully integrated with other hegemonic industries. As history reveals, the rise of industrial agriculture is less the result of a compelling scientific discovery; it is much more the consequence of its linkage to growing a capitalist economy. In Mexico, as in so much of the Americas, industrial agriculture ushered in much more than a technological package to grow seeds. The Green Revolution transformed political power, with the expansion and penetration of capitalism. The Rockefeller Foundation's mission advanced not merely the increased production of commodities; it set in motion a paradigm of production that would eat democracies.

It is for this reason that advancing agroecology depends in large part on gathering a social movement capable of overcoming the entrenched power of industrial agriculture. It is a political terrain that extends beyond dinner tables, shopping baskets, or farms and embraces a much larger social movement to expand political rights, particularly in the United States.

It would be convenient to see the history presented thus far as taking the reader in the direction of what others must do to advance agroecology, particularly campesinos and people living south of the US border. Yet as this history reveals, responsibility for the model of industrial agriculture is derived, in very large part, from US models, institutions, and especially its particular political economy. Indeed, a central feature of the US model of industrial agriculture is its embeddedness in a capitalist economy characterized largely by the absence of democratic governance. Furthermore, upon reflection, one can discern that the political dialogues, discussions about political tactics and strategy, and organizing infrastructures of support and resistance surrounding agroecology are more advanced in many parts of Latin America than is the case of any corresponding effort in the United States.

Consider, for example, the political strategies devised by Latin America's agroecologists to bypass institutions dominated by wealth and power. Campesinos have often relied on "horizontal strategies," as detailed in Chapter 14, to reinforce and build upon relationships between each other, connect neighbouring communities, and establish regional linkages to alternative markets (e.g., tianguis), and social organizations. Such horizontal organizing operates at times as the basis for expanding rights and challenging entrenched power structures. What equivalent political and social organizing strategies exist in the United Stated to advocate for alternatives to industrial agriculture?

Observers typically point to the lack of formal political organization among campesinos as a key barrier to challenging powerful national and global forces. Yet this problem is not unique to campesinos — it is a widespread issue affecting many economic and social groups. This is where a crucial intersection emerges between campesinos and their allies living in cosmopolitan, urban communities across the Americas.

As Peter Rosset has noted, campesino-based organizations like LVC have built important alliances with fisher folk, urban movements, consumers, and others recognizing the need to construct a new social order, beyond the power of industrial agriculture and its alliance with petroleum, chemical, and the finance capital supporting an unrestrained capitalism.[3] Cosmopolitan and urban communities can gain from the experience and understanding of agroecologists with regard to developing alternative markets built on recognizing the value of workers, campesinos, and others. Creating common campaigns both for building supportive infrastructures as well as for resisting oppressive forces (both financial and political) represents a unified strategy adopted by many social movements.[4]

Expanding Spaces for the Exercise of Democratic Rights

There is a straight line running through history from the initial scientific and technical mission in Mexico to the present-day widespread adaptations of science and technology that exposes the erosion of political rights. One of the clearest examples of this begins with food. Narratives of miracle grains and increased yields per acre mask a reality of malnourished populations.

Images of abundant fields belie a world experiencing a cascade of crises on a global scale. Accompanying these signs of collapse, control over food has become increasingly concentrated in the hands of a few powerful corporations. The multiple junctures at which citizens might have exercised political rights over something as basic as nourishment have instead been reduced to consumer choices, determined largely by the class into which one is born.

There is an awakening that the erosion and loss of political rights by the mass of people has been accompanied by the expansion of wealth, rights, and political powers exercised by a vanishingly small class of individuals. This situation defines in large measure a shared purpose for all movements seeking to address one or more of the crises threatening the survival of human civilization: removing the power of an oligarchy and restoring the political rights of ordinary human beings everywhere.

At first glance, science and scientists might seem to play only a peripheral and negligible role in this larger struggle. Yet as this history reveals, communities of scientists have often held a pivotal role in defining problems and their solutions, giving power to different narratives about what is to be done, directing resources and serving as architects of new infrastructures, identifying facts and falsehoods, and perhaps most importantly, exposing relationships: who benefits from technology and who bears the negative consequences.

This history also speaks to the relationship between scientists and social forces. The struggles of Latin America's agroecologists began the moment they recognized that their science was inextricably tied to a powerless class of campesinos. Similarly, the trajectory of many scientists aligned with industrial agriculture was shaped by their ties to a powerful class of industrialists, financiers, and their allies. From these origins, we can witness how two very distinct scientific communities emerged and deployed fundamentally different projects in the world.

Now, the time has come for scientists, and many others, to answer a question voiced by millions of workers around the world at the beginning of the last century:

"Which side are you on?"

Notes

1 The scene described in this chapter is based on a group tour of Cuba organized by Miguel Altieri and Peter Rosset in 1996 and is presented here as recalled by the author.
2 As Rosset and Altieri noted nearly a decade ago, "participating movements warn that agroecology is in danger of being co-opted, given attempts by agribusiness and other actors in the industrial food system to 'greenwash' their discourse, and they reject equating agroecology with industrial monoculture production of 'organic' foods and similar approaches promoted by the private sector and mainstream institutions." See especially "The Dispute for Agroecology" (Rosset and Altieri 2017: 120–5).
3 Peter Rosset, presentation at Food Sovereignty Conference (2015): Peter Rosset at UM Food Sovereignty Conference.
4 Gago and Federici's (2023) presentation provides a broader perspective on the value of campaigns of resistance and their potential for articulating alliances with campesinos to promote a common struggle for such things as food sovereignty.

Epilogue

For my longtime companion, it was just another day at the university. Her schedule included instructing a team-taught program with a fellow faculty member who had published work on the Green Revolution many years earlier. Today's lecture would be led by John, presenting his perspective on that very topic.

The class had barely begun before Professor Cheri Lucas Jennings was on her feet. Facing John across the lecture hall, he seemed visibly unnerved by Cheri's instant willingness to interrupt his lectures. Whereas John thought of the lecture hall as a place to impart his wisdom, Cheri approached teaching as an opportunity for faculty and students to engage in a lively exchange of ideas. Many students recognized a now familiar stance that Cheri reserved for fellow faculty members: she was entirely comfortable with stripping bare their arguments.

"John, pardon my interruption," she began, "but you just stated that Dr. Borlaug as a scientist never intended for his initial work in Mexico to achieve a wider set of purposes beyond increased production on farmers' fields. But what exactly were the metrics he was applying? And weren't those metrics self-limiting, designed to avoid coming to grips with the larger consequences of his work in their fields?" Before allowing John to simply resume his lecture, Cheri continued with the point she wanted students to consider more fully: Was Dr. Borlaug's project a genuine effort to understand the world through scientific inquiry? Or was it, more accurately, an effort to remain wilfully ignorant of the broader consequences?" The look on John's face instantly conveyed to the assembled students that he wished to be anywhere other than the middle of a lecture hall facing Cheri's playful cross-examination.

Cheri often argued that one of the greatest virtues of a truly independent science was for scientists to assume responsibility for the consequences of their work. She would follow this with an important caveat: scientists must acknowledge the political context surrounding their work. That begins with the basic disclosure of sources of funding and support. At a deeper level, she hoped all of her colleagues, scientists and beyond, would engage in a broader discussion framing the nature and applications of their work in the world. Like the conversations among equals (diálogos de saberes) practised by many agroecologists, Cheri had a gift for inviting students and others to share their own forms of knowledge.

She typically urged her students to consider the difference between ethical commitments and the process of inquiry anchored in a democratic process. Among her favourite examples was to compare the avowed commitment by the Green Revolution's scientists to serve humankind versus research designed by agroecologists to illuminate not just ecological and social impacts, but to include campesinos themselves as co-authors of knowledge about the landscapes they worked.

Having taught alongside scientists from many different fields, Cheri was often struck by how so many of her colleagues resisted teaching with those from other disciplines. While always respecting and valuing their expertise, she believed that disciplinary silos too often distanced many scientists from understanding the broader consequences of their work. Claims that they couldn't be responsible for how their science or technology might be used in the world struck her as a form of wilful ignorance. And her critique extended beyond science. Cheri was continually surprised at how often all of us shrink from taking actions, large or small, to address injustices, the abolition of rights and liberties, to not stand up in support of those who are powerless in their struggles against the powerful.[1]

Making change in the world was something that Cheri encouraged not simply among her students, but also among colleagues and friends — and especially in herself. She never asked more of others than she demanded of herself. Having grown up amid the US civil rights movement, Cheri believed that political engagement was quite simply a part of daily life. Large or small, she never missed an opportunity to encourage action, especially in solidarity with those rendered invisible, voiceless, or powerless.

* * * *

A few years later, Cheri and I were preparing a meal during one of our last visits to Kaua'i. It was for Laura, a young lawyer and friend from Berkeley who also happened to be one of Carl Sauer's granddaughters. That evening, our conversation turned to our early life together on Kaua'i and the island's politics.

Being asked to talk about life on Kaua'i, Cheri told one of her favourite stories about a legal case involving Native Hawaiians, their land and water rights, and the late twentieth-century struggles to revive ancient agriculture systems. The conflict, she explained, originated with two large plantations that had long dominated a large portion of the southwestern lands of Kaua'i. Ever since the colonization of Hawai'i at the turn of the century, plantation owners had divided one of the island's principal rivers between themselves to irrigate their sugarcane fields in order to feed a booming export economy.

But as the plantations took more and more of the river's water for their own profits, the river that once had nourished a flourishing Indigenous community began to run dry. In time, conflicts between the two plantations over water rights grew sufficiently adversarial that a legal case was presented to a state court to adjudicate the proper division of water between the two companies. After a lengthy hearing before the court, the judge delivered his decision — one that reportedly caused the legal representatives to fall out of their chairs. Judge Abe concluded that neither plantation held a superior right to the river's water; instead, those having the first claim to all waters of the river were the people of Hawai'i.[2]

For Cheri, that ruling distilled the heart of so many agrarian conflicts in the twentieth century. It was a moment that echoed far beyond Hawai'i, raising the same fundamental questions that defined rural struggles across the world: "Whose land?" or "Whose water?" and gradually evolving into "Whose science?" and "Whose society?"

* * * *

When thinking about solutions, Cheri's disposition was to think broadly and imaginatively. Following the example from the Kaua'i, she explored common legal strategies pursued by Indigenous peoples across the United States to advance agroecological projects in their own communities. Many of these legal efforts, like the Kaua'i case, challenged the water rights from which they had long been dispossessed. These legal strategies shared a common agroecological objective pursued by many other Indigenous people throughout Latin America: food sovereignty.

Cheri also recognized the potential for expanding agroecological projects in other places and other ways. One such example was a proposal led by Miguel Altieri and others in the San Francisco Bay Area to expand agroecological food systems to spaces that included school grounds, university properties, local and regional parks, and a variety of state lands. Projects like this, she believed, offered a powerful opportunity to combine elements to begin building agroecological food systems on a regional basis.

In addition to encouraging students and colleagues to think creatively, Cheri also stressed the importance of thinking critically. For her, critical thinking often began by revisiting historical struggles and engaging with the ongoing work of others pursuing similar goals. These reflections — especially those grounded in diálogos de saberes — provided fertile ground for deeper understanding and more thoughtful solutions.

It would be at this juncture that Cheri would invite us into longer, deeper conversations. She would ask us to pause and reflect on the paths we choose, the consequences they carry, and the cautionary insights offered by those who have been marginalized by past solutions. Such a pause serves as a reminder of what a small band of scientists failed to consider when they gazed upon Mexico's landscapes and devised solutions for others so long ago. At the same time, Cheri tempered reflection with lessons drawn from the historical struggles of many, recognizing "the fierce urgency of now" to achieve social and environmental justice.

In this light, Cheri often questioned some of the most widely promoted alternatives to industrial agriculture. Organic, biodynamic, and regenerative systems each offer clear advantages over conventional models. Yet these approaches can also be viewed as "false solutions," offering limited reforms without addressing the need for a more fundamental transformation, whether it be the power of agribusinesses, oligarchic regimes, or unrestrained capitalism. By 2025, Cheri would no doubt have been alarmed with the emergence of a fascist regime in the United States. I have little doubt that she would have been especially outraged with the demonization of migrant labourers — the vast number of whom were initially forced from their lands decades ago as the result of a scientific and technological mission advanced by powerful economic interests in the United States.

* * * *

Cheri provided a quality that I have endeavoured to weave into the telling of this history, one that I would refer to as conversational. One of the purposes of this history is to encourage a more general conversation about the politics of science, not as a conclusion, but as a starting point for wider public dialogue and political action. Cheri had little patience for authoritative, pedantic assertions that shut down discussion. She believed that meaningful change began with open, ongoing conversations about action.

I confess that Dr. Cheri Lucas Jennings possessed an intellectual depth and breadth that I remain in awe of to this day. While I have tried to capture Cheri's artistry, pedagogy, and intellectual prowess in the making of this book, any shortcomings in this work are necessarily mine alone.

In October of 2020 Cheri Lucas Jennings died at our home in California. This book is but a small part of a much larger contribution she made in the lives of so many, including the many thousands of students who participated in courses she taught at universities in Hawai'i, California, Montana, and Washington. Cheri continues to accompany me as my muse, as a foremost inspiration for thinking and acting politically, as a part of everyday life, and as my love.

Bruce

Notes

1   A regular tool that Cheri used in numerous courses over many decades was a video presentation of a psychological test conducted by Dr. Philip Zimbardo known as the Stanford Prison Experiment. The experiment became notorious for various reasons, including the startling degree to which so many of us may willingly engage and defend acts that are brutal, cruel, and unjust against others.
2   *McBryde Sugar Company, Ltd. v. Robinson* (504 P 2d, 1330, January 10, 1973) No. 4879. A decision rendered by Judge Kazuhisa Abe.

References

Altieri, Miguel. 2004. *Genetic Engineering in Agriculture*. Food First.
———. 2020. "How Bugs Showed Me the Way to Agroecology." *Agroecology and Sustainable Food Systems*, August 24. <bohrium.com/paper-details/how-bugs-showed-me-the-way-to-agroecology/812576099371319296-1543>.
Altieri, Miguel, and Clara Nicholls. 2017. "Agroecology: A Brief Account of Its Origins and Currents of Thought in Latin America." *Agroecology and Sustainable Food Systems* 41, 3–4. <doi.org/10.1080/21683565.2017.1287147>.
———. 2020. "Agroecology and the Reconstruction of a Post-COVID-19 Agriculture." *Journal of Peasant Studies* 47.
Anonymous. 1944. "Mexico Nutrition Studies Progress Report." *International Health Division Annual Report*. Rockefeller Foundation.
———. 1950. "Three Programs and Policies of the Rockefeller Foundation." Rockefeller Foundation Archives, November 29.
Astier, Marta, Jorge Quetzal Argueta, Quetzalcóatl Orozco-Ramírez, et al. 2017. "Back to the Roots: Understanding Current Agroecological Movement, Science, and Practice in Mexico." *Agroecology and Sustainable Food Systems* 41, 3–4.
Banerjee, Neela. 2015. "Exxon's Oil Industry Peers Knew about Climate Dangers in the 1970s." *Inside Climate News*, December 22. <insideclimatenews.org/news/22122015/exxon-mobil-oil-industry-peers-knew-about-climate-change-dangers-1970s-american-petroleum-institute-api-shell-chevron-texaco/>.
Banerjee, Neela, Lisa Song, and David Hasemyer. 2015. "Exxon's Own Research Confirmed Fossil Fuel's Role in Global Warming Decades Ago." *Inside Climate News*, September 16. <insideclimatenews.org/news/16092015/exxons-own-research-confirmed-fossil-fuels-role-in-global-warming/>.
Barbosa, Lia Pinheiro, and Peter Rosset. 2025. *Lessons from the Zapatistas: From Armed Insurrection to Peoples' Autonomy*. Fernwood Publishing.
Bartra, Armando. 1977. "Seis Años de Lucha Campesina." *Revista de Investigación Económica*. Nueva Época: Julio–Septiembre.
Blitzer, Jonathan. 2024. *Everyone Who Is Gone Is Here: The United States, Central America, and the Making of a Crisis*. Penguin Press.
Bradfield, R. 1951. "Memorandum to Warren Weaver, Stakman, Mangelsdorf, and Harrar." Rockefeller Foundation Agriculture, DIMES, June 14.
Bradfield, R., E.C. Stakman, and Paul Mangelsdorf. 1941. "Recommendations of the Commission to Survey Agriculture in Mexico." Rockefeller Foundation Archives, December 3, RG 1.2/323.
———. 1951. "The World Food Problem, Agriculture and the Rockefeller Foundation." Rockefeller Foundation Archives, June 21, RG 3/915/3/23.

———. *Campaigns Against Hunger*. Belknap Press.
Brown, E. Richard. 1979. *Rockefeller Medicine Men: Medicine and Capitalism in America*. University of California Press.
Brown, R. Lester. 1970. *Seeds of Change: The Green Revolution and Development in the 1970s*. Praeger Publishers.
Caire-Perez, Matthew. 2016. A Different Shade of Green: Efraim Hernandez, Chapingo, and Mexico's Green Revolution, 1950–1967. Doctoral dissertation, University of Oklahoma.
Carabias, Julia, and Victor M. Toledo. 1983. *Ecología y Recursos Naturales*. Ediciones PSUM.
Carabias, Julia, Victor M. Toledo, Carlos Toledo, and Cuauhtémoc González-Pacheco. 1989. *La Producción Rural en México: Alternativas Ecológicas*. Fundación Universo Veintiuno.
Carson, Rachel. 1978. *Silent Spring*. Houghton Mifflin.
CGIAR. 1997. "Report of the NGO Committee to the CGIAR Mid-Term Meeting." Cairo, Egypt. <cgspace.cgiar.org/items/a94c6b84-140c-4bed-ac40-e31429628962>.
Chacón, Justin. 2021. *The Border Crossed Us: The Case for Opening the US-Mexico Border*. Haymarket Books.
Chartrand, Vicki. 2019. "Unsettled Times: Indigenous Incarceration and the Links between Colonialism and the Penitentiary in Canada." *Canadian Journal of Criminology and Criminal Justice* 61, 3.
CHC (Canadian Health Coalition). 2018. *Health Accord Break Down: Costs and Consequences of the Failed 2016/17 Negotiations*. Canadian Health Coalition/Ontario Health Coalition.
CIMMYT. 1966–1973. *Annual Reports*. International Maize and Wheat Improvement Center.
CIMMYT. 1974. *The Puebla Project: Seven Years of Experience, 1967–1973*. International Maize and Wheat Improvement Center.
Cleaver, Harry J., Jr. 1975. "The Origins of the Green Revolution." Doctoral dissertation, Stanford University.
Commoner, Barry. 1971. *The Closing Circle: Nature, Man and Technology*. Alfred A. Knopf.
Collinson, M.P., and J.K. Wright Platais. 1991. "Biotechnology and the International Agricultural Research Centers of the CGIAR." Presented at the 21st Conference of the International Association of Agricultural Economists, Tokyo, August 22–29.
Curry, Helen Anne. 2022. *Endangered Maize: Industrial Agriculture and the Crisis of Extinction*. University of California Press.
Dahlsten, Donald, et al. 1970. *Pesticides*. Scientists' Institute for Public Information.
Desmarais, Annette Aurélie. 2007. *La Vía Campesina: Globalization and the Power of Peasants*. Fernwood Publishing.
——— (ed.). 2019. *Frontline Farmers: How the National Farmers Union Resists Agribusiness and Creates Our New Food Future*. Fernwood Publishing.
Dickey, John S. 1951. Memorandum to Warren Weaver, Thomas Parran, William Meyers, and J. George Harrar. Rockefeller Foundation Archives, January, 1.2/323, 2.
Dove, Michael. 1983. "Theories of Swidden Agriculture and the Political Economy of Ignorance." *Agroforestry Systems* 1.

Epstein, Samuel S. 1987. "Are We Losing the War Against Cancer?" *Congressional Record*, Vol. 133, September 9. US Congress.

Ferrell, John A. 1936. "Memorandum to Raymond Fosdick." Rockefeller Foundation Archives, October 16, 323/1.

Fitting, Elizabeth. 2011. *The Struggle for Maize: Campesinos, Workers, and Transgenic Corn in the Mexican Countryside*. Duke University Press.

Fleck, Ludwik. 1981. *Genesis and Development of a Scientific Fact*. University of Chicago Press.

Fossil Fertilizers Report. 2022. *Fossils, Fertilizers, and False Solutions: How Laundering Fossil Fuels in Agrochemicals Puts the Climate and the Planet at Risk*. Center for International Environmental Law. https://www.ciel.org/reports/fossil-fertilizers.

Foucault, Michel. 1982. *The Archaeology of Knowledge*. Vintage Press.

Freire, Paulo. 1970. *Pedagogy of the Oppressed*. Continuum Press.

Gago, Veronica, Cristina Cielo, and Nico Tassi (eds.). 2023. *Economías Populares: Una Cartografía Crítica Latinoamericana*. Consejo Latinoamericano de Ciencias Sociales.

Gago, Veronica, and Silvia Federici. 2023. Presentation. "What Does Re-Enchanting the World Mean Today?" King Juan Carlos of Spain Centre, March 2. <youtube.com/watch?v=KJiNV25dJxg>.

Garcia, Hector Ramos, Catherine Magnon, Juan Manuel Pina, et al. 1984. *La Lucha Campesina en Veracruz, Puebla y Tlaxcala*. Ediciones Nueva Sociología.

Gliessman, Steve. 1990. *Agroecology: Researching the Ecological Basis for Sustainable Agriculture*. Springer.

---. 2016. "Transforming Food Systems with Agroecology." *Agroecology and Sustainable Food Systems* 40, 3. <doi.org/10.1080/21683565.2015.1130765>.

Gonzalez, Juan. 2022. *Harvest of Empire: A History of Latinos in America*. Penguin Books.

GRAIN. 2003. "Sprouting Up — NGO Committee Shuns the CGIAR." Seedling, GRAIN. <grain.org/en/article/359-ngo-committee-shuns-the-cgiar>.

Habermas, Jürgen. 1976. "Theory and Practice in a Scientific Civilization." In Paul Connerton (ed.), *Critical Sociology*. Penguin Books.

Harrar, J. George. 1951. Memorandum to Warren Weaver. Rockefeller Foundation Archives: "Agriculture and the Rockefeller Foundation," DIMES, June 1.

Hernández Xolocotzi, Efraím. 1960. "Pedagogy and Disciplines for Understanding Agriculture in a Socio/Biological Sense." Unpublished conference presentation.

---. 1984. *Las Ciencias Agrícolas y sus Protagonistas*, Vol. 1. Colegio de Postgraduados.

Hewitt de Alcántara, Cynthia. 1976. *Modernizing Mexican Agriculture*. United Nations Research Institute for Social Development.

IPCC (Intergovernmental Panel on Climate Change). 2023. "Summary for Policymakers." In *Climate Change 2023: Synthesis Report. Contribution of Working Groups I, II and III to the Sixth Assessment Report of the IPCC*. <doi.org/10.59327/IPCC/AR6-9789291691647.001>.

Jennings, Bruce H. 1984. "Political Science: A Study of International Agricultural Research." Doctoral dissertation, University of Hawai'i.

---. 1988. *Foundations of International Agricultural Research*. Westview Press.

---. 1990. *California's Experience with Proposition 65: Implementing the Safe Drinking Water and Toxic Enforcement Act*. Senate Office of Research.

———. 1997. "The Killing Fields: Science and Politics at Berkeley, California, USA." *Agriculture and Human Values* 14. <doi.org.10.1023/A:1007469014451>.

———. 2017. *The War on California: Defeating Oil, Oligarchs and the New Tyranny.* Collective Political Strategies.

Jewell Price, Mindy, Alex Latta, Andrew Spring, et al. 2022. "Agroecology in the North: Centering Indigenous Food Sovereignty and Land Stewardship in Agriculture 'Frontiers.'" *Agriculture and Human Values* 39. <link.springer.com/article/10.1007/s10460-022-10312-7>.

Kuhn, Thomas S. 1970. *The Structure of Scientific Revolutions.* University of Chicago Press.

Malkin, Elisabeth. 2009. "NAFTA's Promise Unfulfilled." *New York Times*, April 13, B1.

Martin, O.B. 1921. *The Demonstration Work.* Stratford Co.

McBryde Sugar Company, Ltd. v. Robinson, 504 P.2d 1330 (Haw. 1973), No. 4879. Decision by Judge Kazuhisa Abe, January 10.

Mejido, Manuel. 1972. "Interview with Fernando Carmona." *Excelsior*, November 17. In Roger Burbach and Patricia Flynn (eds.), *Agribusiness in the Americas.* Monthly Review Press.

Mundy, Barbara E. 2018. *La Muerte de Tenochtitlán, la Vida de México.* Libros Grano de Sal.

National Farmers Union. n.d. *Agroecology.* Saskatchewan, Canada. <nfu.ca/learn/agroecology/>.

Nehart, A.J., M.A. Garcia, E. El Hayek, et al. 2025. "Bioaccumulation of Microplastics in Decedent Human Brains." *Nature Medicine.* <doi.org/10.1038/s41591-024-03453-1>.

Noble, David. 1979. *America by Design: Science, Technology and the Rise of Corporate Capitalism.* Oxford University Press.

Oasa, Edmund K. 1981a. "The International Rice Research Institute and the Green Revolution: A Case Study on the Politics of Agricultural Research." Doctoral dissertation, University of Hawai'i.

———. 1981b. "The International Rice Research Institute and the Green Revolution: A Case Study on the Politics of Agricultural Research." University of Hawai'i. <scholarspace.manoa.hawaii.edu/items/coda065c-64d4-4266-852b-8f02727dd6ce>. Citing Andrew Pearse. 1977. "Technology and Peasant Production: Reflections on a Global Study." *Development and Change 8.*

Oreskes, Naomi, and Erik M. Conway. 2011. *Merchants of Doubt.* Bloomsbury Publishing.

Ortencia, Colin Bahena. 1990. "Estudio de la Relación de los Agroecosistemas Frutícolas con la Calidad de Vida Humana en Tetela del Volcán, Morelos, Mexico." Universidad Autónoma del Estado de Morelos, Facultad de Ciencias Biológicas, November.

Otero, Gerardo. 2008. *Food for the Few: Neoliberal Globalism and Biotech.* University of Texas Press.

Padilla, Tanalis. 2008. *Rural Resistance in the Land of Zapata.* Duke University Press.

Paré, Luisa. 1972. "Two Villages in the Puebla Plan: Santa Isabel Tepetzala and Santa Andrés Hueyacatitla." *A Global–2 Report* (unpublished manuscript). United Nations Research in Social Development.

Pearse, Andrew. 1980. *Seeds of Plenty, Seeds of Want: Social and Economic Implications of the Green Revolution.* Oxford University Press.

Peredo Parada, Santiago, and Claudia Barrera-Salas. 2024. "De faros, caravanas y rutas: Estrategias locales para escalonar la agroecología con la gente." Unpublished conference paper, Portugal, September.

Ramos Garcia, Hector, Catherine Magnon, Juan Manuel Pina, Antonio O'Quinn, and Rafael Cardenass Candiani. 1984. *La Lucha Campesina en Veracruz, Puebla y Tlaxcala.* Ediciones Nueva Sociología.

Raphael, Dennis. 2011. *Poverty and Policy in Canada: Implications for Health and Quality of Life.* Canadian Scholars' Press.

Revista del Instituto Nacional Indigenista. 1989. No. 27, Año V. Marzo–Abril.

Rockefeller Foundation. 1959/1960–1964/1965. *Annual Reports.* Rockefeller Foundation. <archive.org/details/the-rockefeller-foundation-annual-report-1959>.

Rockefeller Foundation. 2013. *Food and Prosperity: Balancing Technology and Community in Agriculture.* Rockefeller Foundation.

Rodríguez R., Adriana del Rosario. 2022. Conference presentation: "Agroecología como ciencia transformadora para alcanzar la soberanía alimentaria." Universidad Nacional de Palmira, Colombia. YouTube@UNALsede.Palmira.

Rosenberg, Charles E. 1976. *No Other Gods: On Science and American Social Thought.* John Hopkins University Press.

Rosset, Peter M. 1999. *The Multiple Functions and Benefits of Small Farm Agriculture.* Food First Policy Brief #4. Institute for Food and Development Policy.

Rosset, Peter, and Miguel Altieri. 2017. *Agroecology: Science and Politics.* Fernwood Publishing. Citing La Vía Campesina. 2015. "Declaration of the International Forum for Agroecology."

Rosset, Peter, Ivanete Ferreira Fernandes, Lia Pinheiro Barbosa, et al. 2025. "Unlearning the Green Revolution: Inventory of Agroecological Practices in Ceará, Brazil, an Instrument for Decolonizing Territory and (Re)Valuing Peasant Knowledge." *Environmental Science and Policy* 165. <doi.org/10.1016/j.envsci.2025.104022>.

Rosset, Peter, Lia Pinheiro Barbosa, Valentín Val, and Nils McCune. 2020. "Pensamiento Latinoamericano Agroecológico: "The Emergence of a Critical Latin American Agroecology?" *Agroecology and Sustainable Food Systems* 44.

———. 2022. "Critical Latin American Agroecology as a Regionalism from Below." *Globalizations* 19, 4.

Rubio, Blanca. 1987. *Resistencia Campesina y Explotación Rural en México.* Ediciones Era.

Rushe, Dominic. 2022. "Big Agriculture Warns Farming Must Change or Risk Destroying the Planet." *The Guardian*, November 3. <theguardian.com/environment/2022/nov/03/big-agriculture-climate-crisis-cop27>.

Sanchez, Juan. 1992. "Recuento de Actividades de Captación del CLADES." *Revista del Consorcio Latinoamericano sobre Agroecología y Desarrollo* 4.

Sauer, Carl O. 1945. "Letter to Joseph Willits." Rockefeller Foundation Archives, February 12, 1.2/323.

———. n.d. "Memo Regarding Wallace's Idea for a Program in Mexico." Rockefeller Archive Center, Rockefeller Foundation Records, Projects, RG 1.2, Series 323, Box 10, Folder 63.

Secretary of State, California. n.d. "History of Initiatives Voted into Law." <sos.ca.gov/elections/ballot-measures/resources-and-historical-information/history-california-initiatives>

Shiva, Vandana. 2016a. *Who Really Feeds the World? The Failures of Agribusiness and the Promise of Agroecology.* North Atlantic Books.

———. 2016b. *The Violence of the Green Revolution: Third World Agriculture, Ecology and Politics.* University Press of Kentucky.

Siebert, Steve, and Jill Belsky. 2014. "Historic Livelihoods and Land Uses as Ecological Disturbances and Their Role in Enhancing Biodiversity: An Example from Bhutan." *Biological Conservation* 177. <sciencedirect.com/science/article/abs/pii/S0006320714002420>.

Strategy for Small Farmer Development. 1975. *An Empirical Study of Rural Development Projects.* Development Alternatives, Inc.

Taussig, Michael. 1978. "Peasant Economics and the Development of Capitalist Agriculture in the Cauca Valley, Colombia." *Latin American Perspectives* 5, 3.

Tickner, Joel. 2003. *Precaution: Environmental Science and Preventive Public Policy.* Island Press.

Toledo, Victor M., and Narciso Barrera-Bassols. 2009. *La Memoria Biocultural: La Importancia Ecológica de las Sabidurías Tradicionales.* ICARIA Editorial.

———. 2017. "Political Agroecology in Mexico: A Path toward Sustainability." *Sustainability*, February 14.

Trujillo Arriaga, Javier. 1990. "Biotecnología o agroecología: Selección de paradigma tecnológico para el desarrollo campesino en México." In Blanca Suárez (ed.), *Biotecnología para el Progreso de México.* Centro de Ecodesarrollo.

UNAL (Universidad Nacional de Colombia, Palmira). 2022. "Agroecología como ciencia transformadora para alcanzar la soberanía alimentaria." Presentation by Adriana del Rosario Rodríguez R. YouTube@UNALsede.Palmira.

US Centers for Disease Control and Prevention. 2024. *National Health and Nutrition Examination Survey NHANES Biospecimen Program.* <cdc.gov/nchs/nhanes/biospecimen/index.html>.

US Office of Technology Assessment. 1987. *Technologies to Maintain Biological Diversity.* US Congress.

Val, Valentín, Peter M. Rosset, Carla Zamora Lomelí, Omar Felipe Giraldo, and Dianne Rocheleau. 2019. "Agroecology and La Vía Campesina I: The Symbolic and Material Construction of Agroecology through the Dispositive of 'Peasant-to-Peasant' Processes." *Agroecology and Sustainable Food Systems* 43, 7–8. <doi.org/10.1080/21683565.2019.1600099>.

van den Bosch, Robert. 1978. *The Pesticide Conspiracy.* Doubleday.

Weaver, Warren. 1950. "Inter-Office Correspondence to Chester I. Barnard." Rockefeller Foundation Archives, September 21, 1.

———. n.d. "Weaver Memorandum to John S. Dickey, Thomas Parran, and William Meyers." Rockefeller Foundation Archives 1.2/RG 323.

Wellhausen, Edwin J. 1956. "Memorandum to J. George Harrar." Rockefeller Foundation Archives, August 18. Agri Minutes 323, box (folder not recorded).

———. 1961. "Mexican Agricultural Program." Rockefeller Foundation Archives, October 2, 323/2/10.

Worster, Donald. 1977. *Nature's Economy: A History of Ecological Ideas.* Cambridge University Press.
———. 1979. *Dust Bowl: The Southern Plains in the 1930s.* Oxford University Press
———. 1985. *Rivers of Empire: Water, Aridity, and the American West.* Oxford University Press.
Wright, Angus. 1991. *The Death of Ramón González.* University of Texas Press.

The Struggle Continues......

The struggle to advance a revolutionary science builds on the efforts of countless others already committed to expanding agroecology across the globe. We believe that strengthening the scientific and social movement for agroecology is essential to the livelihoods of millions of campesinos, the health of our planet, and the well-being of current and future generations.

There are a variety of ways to encourage and support agroecology via social media, educational forums, alliances, supportive infrastructures such as food sovereignty, and community-based projects. What matters most is participating at whatever level you can, while remaining mindful of the need to build collective, mass-based efforts that advance popular action.

To learn more about how you can join many others to advance this movement, please visit us at https://www.brucehjennings.com.

Index

agribusiness, 4, 10, 37, 45, 48, 50, 53, 61, 65–66, 85, 92, 106, 128–130, 144
agricultural,
   chemicals, 60, 68, 71, 76–77, 80
   ecology, 67, 95
   economy, 22, 39, 55
   extension, 51
   landscape, 11, 57, 66
   model, 67
   plagues, 59
   production techniques, 121
   related institutions, 36
   research, 3–4, 35, 37, 46, 50, 68, 82, 107, 121, 137
   technicians, 37
agroecological,
   engineering, 95
   knowledge, 67
   lighthouses, 122
ahupuaʻa, 5n1
ʻāina, 2
alternative markets, 120, 131, 138
Altieri, Miguel A., 3–4, 66, 88, 92, 95, 98, 107, 140n1
American Petroleum Institute (API), 83
Anderson Clayton, 37
Article 27 of Mexico's Constitution, 62

Bartra, Armando, 113
Bartra, Roger, 59
Big Green (Proposition 128), 82–85
biological control, 68–73, 121. *See also* Division of Biological Control
Borlaug, Norman, 16, 62
Bosch, Robert van den, 70
botany, 34, 36, 39, 41
Bradfield, Richard, Dr., 16–17, 21, 23–24, 46–48
Brazil, 48–49, 106

California,
   agricultural fields, 60
   Central Valley, 5, 75
   initiative, 81
   Legislature, 4
Camacho, Manuel Ávila, 23
campesino(s),
   based movements, 56
   centred agriculture, 45
   centred program, 40
   household, 26, 31, 56
   and their communities, 28, 52, 97, 100, 108
Canada, 3, 102, 105, 130
cancer, 80–82, 91
capital, 9, 23, 29, 39, 43, 47, 97, 106, 112, 127, 138
capital intensive inputs, 122, 133
capital-intensive agriculture, 112
capital-intensive infrastructure, 39
capitalist, 14, 20, 37, 51, 59–60, 62, 90, 105, 137
   broader critique of capitalism, 55
   farming, 9–10, 23, 55, 59, 69–71, 96–97, 107, 116, 122–123, 133–134
   logic, 54, 61–62, 105
Carabias, Julia, 92
carbon recycling, 123
Cárdenas, Lázaro, 22
Carson, Rachel, Dr., 69
   Silent Spring, 69, 77
Cauca Valley, 59
Chapela, Ignacio, 114
Chávez, César, 73
Chile, 5, 25, 48–49, 66–67, 106, 122–125, 129–130
   Chiloé, 5, 124–125
   Yumbel, 122–123, 131
chinampas, 11, 88, 93

circular versus linear models of farming, 122
classical biological controls, 69, 121
climate change, 15n3, 83
Colegio de la Frontera Sur (ECOSUR), 96, 102, 116
Colombia, 46–49, 58–59, 66–67, 99, 106, 125–126
  Cauca Valley, 59
colonization of knowledge, 50–51
commercial agriculture, 24, 50
commercial crops, 60
Commoner, Barry, 70, 72n1
  The Closing Circle, 72
communal and ejido property rights, 49
communal lands, 62
Consultative Group on International Agricultural Research (CGIAR), 50, 52, 97, 107–108, 121
contaminants in food, 65, 111, 135
contracted labor, 59–60
conventional agriculture, 69, 89, 95–96
corn, 16, 18, 25, 29, 105–106, 112, 114
Cornell University, 23, 34
cotton, 17, 35, 60, 92
Counterrevolution, 7, 14, 27, 29, 31, 33, 49, 70, 111, 113, 136
creative destruction, 60–61, 105–106
Curry, Helen Anne, 63n1

Dahlsten, Donald, Dr., 70, 110–111, 115, 117n1
de Alcántara, Cynthia Hewitt, 59
democratic rights, 104, 130, 138
dialogo de saberes, 126
Dickey, John S., Dr., 30
Division of Biological Control, 68–70, 73, 110–111, 113, 115, 117n2
Dove, Michael, 134
Dow Chemical, 37

ecological,
  destruction, 42
  effects, 81
  engineering, 88
  hazards of pesticides, 73
  impacts, 66, 121
  resources, 60

ecology, 5n1, 11–12, 25, 34–36, 41, 49–52, 66, 68, 71–72, 95, 98, 128
economic,
  politics, 107
  pyramid, 98
  value, 71
ejidos, 22, 62, 125
elite farmers, 35, 45, 62
entomology, 67
environmental,
  consequences, 59, 81
  damage, 97
  health, 79n1, 122
  laws, 82
Environmental Protection Agency, 79n1
epistemological innovations, 99
ethnobotany, 49, 99
extensionism, 41, 108, 126
Exxon corporation, 83

Farmer First, 97
Farming Systems Research, 97
farmworkers, 71, 74
fellowships, 48–49, 97
Ferrell, John, Dr., 17, 22
fertilizers, 18, 29, 36, 40, 43, 122, 133, 135
finance, 22, 37, 50, 52, 59, 61–62, 104, 129, 138
financial,
  elites, 57
  support, 52
  transactions, 74
fish ponds, 1, 11
Fitting, Elizabeth, 117n3
Fleck, Ludwik, 14n2, 44n3
floating gardens (chinampas), 11, 88, 93
food,
  access, 128
  industry, 40
  safety, 82
  sovereignty, 120, 122, 126–129, 132n1, 136–137, 140, 140n3, 140n4
  supplies, 38
Ford foundation, 56, 66, 97
forestry, 90, 125
Fosdick, Raymond, 16–17
fossil fuel, 13, 76, 83–84, 135–136
Foucault, Michel, 44n3

free market, 43, 62, 74, 83, 105
  economy, 37
  forces, 62
free trade, 61, 97, 102–105, 107
Freire, Paulo, 44n3

Gago and Federici, 140n4
Gates foundation, 121
General Education Board (GEB), 17, 21
General Union of Mexican Workers and Peasants (UGOCM), 54
genetic engineering technologies, 108
genetics, 10, 23, 41
Gliessman, Steve, 92, 95
globalization, 112
GM corn, 112, 114
Green Revolution, 18, 45, 51, 52n2, 57, 60–61, 63, 68, 93, 97–98, 106–107, 109, 125–126
  institutional relationships, 20, 22, 50
  landscapes, 53
  mentality, 90
  revolutionaries, 69
  solutions, 18, 50, 89, 96, 115, 139
green revolutionaries, 69
greenhouse gases, 10, 82, 84
Guatemala, 5, 48–49, 106

hacendados, 34, 104
Hanakāpī'ai, 1–2
Harrar, J. George, 47
Harvard University, 16, 23, 34
harvesting, 36, 45, 135
health hazard, 78, 79n1
Herrera, Oscar Brauer, 39
Honduras, 48–49, 106
horticultural-fruit production systems, 66
hunger, 10, 18, 42, 58
hybrid seed development, 41, 43

improved seeds, 52, 59
Independent Federation of Agricultural Workers and Peasants (FIOAC), 91
Indigenous,
  agriculture, 11
  communities, 57, 116
  heritage, 44n2
  people, 3, 51, 105, 130

industrial,
  agricultural scientists, 66, 102
  agriculture, 9–14, 15n3, 44, 48, 68, 74, 97, 134–139
  food production model, 132
  trees, 124
industrialization, 108, 112–113, 115
Infante, Augustin, 122
infrastructure, 16, 39, 52, 62, 116, 119–121, 124–129, 131, 133, 139
Institute for Agricultural Investigations (IIA), 20, 28–29, 40, 45
International,
  trade, 62, 103–105
  trade treaty, 62
  international agricultural research centres (IARCs), 3–4, 46, 50–52, 68, 96–97, 101, 107–108, 137
International Center for Tropical Agriculture in Colombia (CIAT), 50, 67
International Health Division of the foundation, 17
International Maize and Wheat Improvement Center (CIMMYT), 35, 46, 49–50, 55–56, 63, 67, 97–98, 101–102, 107, 109, 112, 121
  Economics Program, 97
  Puebla Program, 97
International Rice Research Institute (IRRI), 46, 97–98
Iowa, 25–26, 29–31
irrigation, 1, 18, 36, 43, 52, 59, 90, 122

Jaramillistas, 55
Jaramillo, Ruben, 55 Jennings, Cheri Lucas, 1, 8, 88

Kaua'i case, 143
killing fields, the, 106
Knapp, Seaman, 17, 62
Kuhn, Thomas S., 14n2, 44n3

La Central Independiente de Obreros Agrícolas y Campesinos (CIOAC), 92
La Escuela de Ciencias del Mar de la Universidad Autónoma de Sinaloa, 44n1, 92

La Universidad Autónoma de Chapingo
  (UACH), 44, 44n1, 94, 109
La Universidad Nacional de Colombia,
  126
La Vía Campesina (LVC), 105, 116, 127,
  130, 132, 137–138
labour,
  conflict, 60
  force, 112
  organization, 59
  organizers, 58, 116
  rights, 48
  unrest, 60
land,
  access rights, 55
  grant colleges, 17, 68–70
  poverty, 58
  rights, 121
  sovereignty, 130
  titles, 48
latifundios, 22
Latin American Consortium on
  Agroecology and Development
  (CLADES), 98
Latin American Institutes of Agroecology
  (IALAs), 127
Latin American Scientific Society of
  Agroecology (SOCLA), 99–100
Lia Pinheiro Barbosa and Peter Rosset,
  106
Lessons from the Zapatistas, 105–106

machinery, 29, 36, 40, 47, 71, 90
maize, 23–25, 35, 46, 55–56, 93, 113–115,
  120–121
Manglesdorf, Paul, Dr., 16–17, 23
marginalized communities, 51
market,
  access, 121
  cooperatives, 131
  forces, 54, 60, 62, 75
  instability, 60
  regimes, 52, 54, 125
Mexican Agricultural Program (MAP),
  25–26, 30, 34, 43, 45–48, 59
Mexico,
  Baja California, 54, 92
  Chiapas, 57, 102–105, 116

Chihuahua, 55
Ciudad Obregon, 92
Durango, 35
Guerrero, 57
Hidalgo, 57
Irapuato, 92
La Comarca Lagunera, 91
Mazatlan, 91
Morelos, 55, 93
Oaxaca, 57, 114
Puebla, 55–57, 63, 97
San Cristobal, 103–105
Sinaloa, 54, 91–92
Sonora, 37, 54, 92
Tlaxcala, 57
Veracruz, 57
Xochimilco, 5, 88
Yaqui, 92
Yucatan, 92, 102
Mexico's Ministry of Agriculture, 20
Mexico's national health department, 22
migrant agricultural workers, 73
monoculture, 10, 51, 67, 69–70, 93, 115, 122,
  140n2

national
  departments of agriculture, 48
  labour laws, 73
  peasant union, 57
National Farmers Union (NFU), 3, 103,
  130
National Institute for Agricultural
  Investigations (INIA), 45
nationalization, 60
native maize (criollos), 113–114
natural resources, 35, 52, 90, 107, 125
neutral science, 31, 58, 61
Nicaragua, 49, 106
niche markets, 75
non-irrigated regions, 29
North American Free Trade Agreement
  (NAFTA), 102–106, 112
nutrition, 17, 68, 127–128

Oasa, Edmund K., Dr., 98
Office of Environmental Health Hazard
  Assessment, 79n1   Office of Special
  Studies (OEE), 20, 29, 34–35, 38, 45

organic, 75, 81, 95, 140n2
  systems, 144

Palomino, Ramon Danzos, 55
Paré, Luisa, 57, 59
patronatos, 37, 43, 57, 62
Pearse, Andrew, 59
Peru, 5, 25, 48–49, 106
  the valleys surrounding Cuzco, 5
pest management, 89, 91, 96
The Pesticide Conspiracy, 70
Pesticides,
  consequences on human health, 77
  exposure, 73–74, 78
  exposures to harmful residues on fruits and vegetables, 60
  hazards, 69, 77
  injuring and killing fieldworkers, 60
  pesticide-dependent monoculture model, 51
  pesticide-free, 75
  poisoned soils, 60
  polluted drinking water, 60
  toxic air contaminants, 60, 135
petrochemical corporation, 65
petroleum companies, 72, 76, 84
Pioneer Hi-Bred, 37
Plan Puebla, 55–56, 63
plant breeders, 24, 46
plant production, 37
poisoned waterways, 121
poisoning of workers and their communities, 111
political,
  bosses (caciques), 55–57
  conflicts, 48
  connection, 55
  debate, 74
  implications, 34
  politics of science, 3, 110, 121
  power, 13, 42, 71, 127–128, 137, 139
  relationships, 50
pollution and environmental damages, 97
polycultures, 11, 93, 120
process of creative destruction, 106
processed foods, 121, 135
production,
  alternatives, 101

paradigm, 53
practices, 57, 75, 80, 121, 136
production/consumption systems, 51
productivity, 18, 20, 42–43, 53, 113, 128
Proposition 65, 80–81
public health, 8, 22, 53, 62, 73–78, 81, 99, 121

Quist, David, 114

Red Mexicana de Tianguis, 131
redesigned crop, 36
reductionism/reductionist, 41–42, 49, 76–77, 111, 114–115
regenerative systems, 144
regional economy, 50, 52, 56, 98, 122
reproductive harm, 80, 82
residues in food, 76
resilient systems, 52
Rivera, Diego, 27, 33
Rockefeller foundation, 7, 13–14, 16, 22–23, 26–28, 34, 45, 55–56, 58, 112
Rockefeller, John D., 13–14, 21, 53
Rockefeller model, 29
Rockefeller philanthropy, 17
Rodriguez, Adriana del Rosario, 125
Rojas, Rolando, 123
Rosset, Peter, 4, 50, 102, 106, 116, 126–127, 132n1, 138, 140n1, 140n3
Ruiz, Archbishop Samuel, 103
rural,
  conflicts, 57, 107
  households, 53, 56, 87
  peace, 55
  resistance, 58
  violence, 61

Safe Drinking Water and Toxic Enforcement Act, 80
Saldana, Tomas Martinez, 109
Sauer, Carl, Dr., 24
Schultes, Richard, 16
scientific,
  agriculture, 13–14, 18–21, 28–29, 36, 46–50, 54, 56–58, 89
  agriculture model, 89
  reductionism, 42, 77
seed exchange networks, 131

Shell, 37, 83
Shiva, Vandana, 59, 63n3
Silent Spring, 69, 77
silo effect, 120
single crop (monoculture), 10, 51, 69–70, 93, 115, 122, 140n2
social,
    conflict, 56–58, 61
    consequences, 30, 32, 53, 57, 77
    scientists, 41, 53, 97, 106
    social-bound problems, 56
    socioeconomic disruptions, 56
    socioeconomic impacts, 56, 109
    sociological studies, 38
    sociology, 41, 44n3, 49
    tensions, 31, 42, 53, 55, 59
    welfare, 61
    social movements, 96, 116, 121, 137–138
soils, 23, 60, 90, 93, 111, 122–123, 130
species extinction, 121
Stakman, Elvin Charles, Dr., 16, 23
structural violence, 105–106
sustainable food systems, 96
synthetic fertilizers, 122, 133
synthetic organic chemicals, 81
synthetic pesticides, 68–69, 71, 75, 77, 83

Taussig, Michael, 59–60
technical assistance, 16, 19, 23–25, 28, 32–33, 39, 42, 47, 51, 55
técnicos, 37, 48, 50–51, 59
territorialization, 126
Texaco, 83
tianguis, 131, 138
Toledo, Victor M., 93
Traditional Agricultural Systems,
    cacatales, 93
    cafetales bajo sombra, 93
    campos drenados, 93
    campos elevados, 93
    chinampas, 11, 88, 93
    cultivos de escorrentía, 93
    huertos familiares, 93
    marceno, 93
    pasto marino en la costa, 93
    policultivos sobre dunas, 93
    terrazas y policultivos, 93
traditional agriculture, 44

traditional knowledge, 4, 11, 51–52, 93, 96
Tratado de Libre Comercio (TLC), 102–104

United Farm Workers union, 73
United Fruit, 59
United Nations agencies, 46
United States, 3, 48, 50, 53, 72, 89–90, 92, 105–106, 112, 127, 136–138, 143
Universidad Católica Boliviana, 126
University of Minnesota, 16, 23
urban agriculture, 128
urban gardens, 131, 133
US Agency for International Development, 52
US agricultural scientists, 16, 24, 43, 98

village schools, 120, 131

waru-waru, 11
Weaver, Warren, 30, 32
Wellhausen, Edwin, Dr., 25, 37–38, 45
western epistemologies, 102
wheat, 16, 18, 23, 25, 35, 46, 121, 123
working conditions, 60, 73, 107
workplace rights, 76
World Health Organization (WHO), 91
worsening diets, 97
Worster, Donald, 8, 14n1

Xolocotzi, Efraím Hernández, 34–36, 38–43, 44n2, 61, 92

Zapata, Emiliano, 55
Zapatistas, 105–106